KILLER ROBOTS

Lethal Autonomous Weapon Systems
Legal, Ethical and Moral Challenges

KILLER ROBOTS

Lethal Autonomous Weapon Systems
Legal, Ethical and Moral Challenges

by

Wing Commander (Dr) U C Jha

Vij Books India Pvt Ltd

New Delhi (India)

Published by

Vij Books India Pvt Ltd
(Publishers, Distributors & Importers)
2/19, Ansari Road
Delhi – 110 002
Phones: 91-11-43596460, 91-11-47340674
Fax: 91-11-47340674
e-mail: vijbooks@rediffmail.com

Contents

Preface   vii

List of Abbreviations   ix

Chapters

I   Introduction   1

II   Lethal Autonomous Weapon Systems: Definition   13

III   Autonomous Weapons in Use   30

IV   Lethal Autonomous Weapon Systems (LAWS)
        and International Law   69

        1. Compliance with International Humanitarian Law   69

        2. Violation of International Human Rights Law   86

V   LAWS: Ethical and Moral Issues   105

VI   LAWS: Legal Review   122

VII   LAWS: International Concerns   148

VIII   Conclusion and the Way Ahead   177

Appendices

A   Relevant International Humanitarian Law Provisions   195

B   Model CCW Protocol for Regulation of LAWS   212

Bibliography   215

Source of Photographs   242

Index   243

Preface

Killer robots or lethal autonomous weapon systems (LAWS) may in the not-so-distant future be able to select targets and use lethal force without human intervention. Though LAWS in the true sense are not yet available, at least 44 countries, including China, France, Germany, India, Israel, South Korea, Russia, the UK and the United States, are developing such capabilities. The existing weapons that could be categorized as autonomous are used only in conflict zones and the types of targets they can attack, and the circumstances in which they are used are pre-determined. In contrast to LAWS, the existing autonomous weapon systems are overseen in real-time by a human operator.

The proponents of LAWS argue that such systems would be capable of meeting the laws of war or international humanitarian law (IHL) standards. Those against the development of LAWS argue that such weapons would violate the basic principles of IHL and threaten the right to life and principle of preserving human dignity. The proliferation of LAWS could have a destabilizing effect on international security. They could also fall into the hands of non-state actors. To say that future LAWS would be more accurate, IHL abiding and reduce the suffering of those out of armed conflict seems like wishful thinking when one considers the fact that the use of the most advanced weaponry in the conflict in Syria has not reduced the suffering of innocent civilians. Once available in the military arsenal, LAWS would be perfect tools of repression in the hands of a few developed States.

In July 2015, nearly 1,000 eminent scientists, robotic professionals, and international law experts released a statement promoting a ban on LAWS. A number of states, prominent world leaders, the Secretary-General of the United Nations, and the International Committee of the Red Cross (ICRC) have expressed serious concerns about the development of LAWS. Diplomats and experts held discussions on the legality of LAWS at the Palais des Nations in Geneva in April 2016. The International Committee for Robot Arms Control (ICRAC) has urged the international community to seriously consider the prohibition of autonomous weapons systems in the light of the dangers they pose to global peace and security. The Human Rights Watch and Harvard Law School's International Human Rights Clinic have also urged the states to ban such weapons over which humans have no control.

The present work offers an overview of the existing autonomous weapons. It evaluates the moral and legal issues relating to the development and the deployment of LAWS in armed conflict and analyses international concerns over such a possibility. It also makes recommendations for a pre-emptive ban on the development of such weapons. The work would be successful if it contributes to a better understanding of the consequences of the development of LAWS and hence, the need for a pre-emptive ban because once such weapons are inducted, the States would be unwilling to discard them.

I am thankful to my wife, Ratna for her constant support, and to Medha Dubey and Chandana Banerjee for editorial assistance. This book would not have been possible without the cooperation of Brigadier P K Vij (Retd) who was instrumental in its publication.

— U C Jha

List of Abbreviations

AAR	Autonomous Aerial Refuelling
A-CTUV	Anti-Submarine Warfare (AWS) Continuous Trail Unmanned Vessel
ADF	Australian Defence Force
AP	Additional Protocol
ATT	Arms Trade Treaty
AWS	Autonomous Weapons Systems
CCW	UN Convention on Prohibitions or Restrictions on the Use of Certain Conventional Weapons
DARPA	Defence Advanced Research Projects Agency
DMZ	Demilitarized Zone
DoS	Denial of Service
DRDO	Defence Research and Development Organization
EOD	Explosive Ordnance Disposal
ERW	Explosive Remnants of War
FLIR	Forward-Looking InfraRed
GC	Geneva Convention
HMMWV	High-Mobility, Multipurpose Wheeled Vehicle
HRC	Human Rights Committee
IAI	Israel Aerospace Industries Limited
ICC	International Criminal Court
ICJ	International Court of Justice
ICCPR	International Covenant on Civil and Political Rights

ICRAC	International Committee for Robot Arms Control
ICRC	International Committee of the Red Cross
IED	Improvised Explosive Devices
IHL	International Humanitarian Law
ISR	Intelligence, Surveillance and Reconnaissance
LAR	Lethal Autonomous Robot
LAWS	Lethal Autonomous Weapon Systems
LDUUV	Large Displacement Unmanned Undersea Vehicle
LOAC	Law of Armed Conflict
LOCAAS	Low Cost Autonomous Ammunition System
LRASM	Long-Range Anti-Ship Missile
LS3	Legged Squat Support System
MAARS	Modular Advanced Armed Robotic System
MHC	Meaningful Human Control
NGO	Non-governmental organization
PRIO	Peace Research Institute Oslo
ROE	Rules of Engagement
SAWS	Semi-autonomous Weapon Systems
SIPRI	Stockholm International Peace Research Institute
SWORDS	Special Weapons Observation Remote Direct-Action System
TRADOC	The US Army's Training and Doctrine Command
TUGV	Tactical Unmanned Ground Vehicle
UAS	Unmanned aircraft systems
UAV	Unmanned Aerial Vehicle
UCAV	Unmanned Combat Aerial Vehicle
UGS	Unmanned Ground System
UGV	Unmanned Ground Vehicle
UMS	Unmanned System
UN	United Nations

UNIDIR	United Nations Institute for Disarmament Research
UNHRC	UN Human Rights Council
UNROCA	UN Register of Conventional Arms
UNSC UN	Security Council
USDoD	United States Department of Defence
USV	Unmanned Surface Vehicle
USSV	Unmanned Surface Sea Vehicle
UUV	Unmanned Underwater Vehicle
UV	Unmanned Vehicle

I Introduction

Imagine a Robotic-Sepoy[1] (Roby) monitoring the movement of the enemy from across the international border. Roby is five feet tall, can travel 5 km an hour, weighs 450 kg and can spin on a dime. It has been programmed to select and engage targets without intervention by a human operator. Roby can track multiple moving targets using IR and visible light cameras and 'see' up to a distance of 2000 metres at night, give a warning to an intruder and if not obeyed, fire a lethal shot with a deadly accuracy when the intruder is within a range of 250-300 metres. It can also jam radios and cell phones nearby. Roby is not alone; it is one of the many armed robots positioned at the international border to serve as the first line of defence.

Roby is an autonomous weapon system. It is unaffected by the emotions and stress that cause soldiers to overreact or deliberately disobey rules of engagement and commit war crimes. Roby will be on duty for 24 hours, 365 days under all weather conditions. It will not ask for any leave, not go to sleep and not require food. It will only require regular charging and if nuclear-powered, it would remain active for years without any recharge. It will also not require any expenditure on healthcare, accommodation, or entertainment, nor will it ask for pension on retirement. What is more, it will not pose disciplinary problems such as desertion, disobedience and fratricide. In a number of countries, militaries are planning for a day when sentry robots like Roby will stand guard at borders, ready to identify the enemy and kill them, without an order from a human soldier.

Evolution of Robotics

Though man has been obsessed with robots[2] or automation for a long time, the large-scale use of robots has become possible only recently with progress in areas such as computation, data storage, the Internet, wireless communication, electronics and design and manufacturing tools.[3] Today, robotic functions are a part of many household equipment or gadgets used for cleaning, mowing the lawn, cooking, serving, counting, and nursing patients. High-end cars have various features for autonomous functions such as anti-lock brakes, traction and stability control, power steering, emergency seat belt retractors, mechanism for avoiding collisions, and air bags. Several manufacturers and research organizations have even produced robotic or driverless cars and these are being tested in many countries. Google's self-driving car, for example, had driven close to two million miles with only 11 minor accidents, most of them caused by human errors.[4] In the wake of anthrax scares, robots are increasingly being used in postal sorting applications. The US postal service has estimated that it has the potential to use up to 80,000 robots for sorting.

In the industrial sector, robotic machines are being used to carry out various tasks that were earlier performed by people.[5] In military aircraft, automatic ground collision avoidance systems can take control of a human-piloted aircraft if a pilot becomes disoriented. Commercial airliners have a high degree of automation for every phase of a flight.

We are now poised to have robots fight for us. When the US forces went into Iraq in 1991, there were no robotic systems on the ground. By the end of 2004, there were 150 robots on the ground in Iraq; a year later there were 2,400; and by the end of 2008, there were about 12,000 robots of nearly two dozen varieties. Autonomous robots have also been deployed at the Israeli-Palestinian border and the Korean Demilitarized Zone. In Iraq and Afghanistan, robots were deployed mainly in dull, dirty or dangerous tasks such as disrupting or exploding improvised explosive devices and surveillance in dangerous areas such as caves. The vast majority of the US ground robots were used to explode improvised explosive devices. Currently, a human operator remains in the loop for decision-making regarding the deployment of lethal force; however, future systems are likely to be completely autonomous. They may be land-based or air-based and may operate on or under the surface of water. Autonomous systems are also being designed to carry load for soldiers travelling on foot in extreme terrain and to rescue the wounded.

Robotic Weapon Systems

A few countries have already produced and deployed autonomous military systems. The Israel Aerospace Industries Limited (IAI) has produced the operational autonomous weapon Harpy, which is a lethal anti-radar attack system. It loiters over the battlefield, detects and identifies radar emitters, and flies into targets to destroy them. The US Navy's MK 15 Phalanx Close-In Weapons System is capable of autonomously performing search, detection, evaluation, track, engage and kill functions. The US Navy's X-47B is a fully autonomous aircraft that has been successfully launched from and landed on an undocked aircraft carrier. It is currently unarmed; its design enables it to support up to a 4,500-pound payload in its twin internal weapons bays. For the US, a robot is a cheaper option than a soldier. A soldier costs the US about $4 million during his/her lifetime. A robot would cost less than 10 per cent of that and a robot can be scrapped when it is damaged or obsolete.[6]

South Korea has deployed the SGR-A1 in the Korean Demilitarized Zone. This land-based system is designed to select and track targets automatically and is capable of making the decision to fire at a target, completely independent of human input. The UK has developed the Taranis aircraft, which is capable of autonomous flight, but it has not yet been weaponized. The British Parliament has announced that the UK is not developing and will not be using fully autonomous weapon systems. The Indian Defence Research and Development Organization (DRDO) has indicated that it is in the process of developing robotic soldiers and mules for the armed forces.

It is very likely that lethal autonomous weapon systems (LAWS) developed in the next two decades will have the capability to hunt, identify, authenticate and possibly kill a target without a human in the decision loop. Lethal robots could also be used as force multipliers in the battlefield. It is claimed by the proponents of such weapon systems that they would be capable of adhering to the rules of international humanitarian law (IHL) and would, in fact, perform more ethically than soldiers in the battlefield.[7]

Advantages

The supporters of LAWS are of the strong opinion that substituting humans with machines in armed conflict would not only save lives, but improve performance. Robots would be able to carry out their missions for extended periods of time and would lighten the burden of soldiers by improving their physical security and performing logistic support tasks.[8] The perceived advantages of LAWS are the following:

- They would provide protection to soldiers by reducing exposure to risks, for instance, by assisting in reconnaissance.

- They would process battlefield information faster and more efficiently than humans.

- They could be deployed at a much greater speed than human soldiers. This would help in immediate reaction to a threat with greater efficiency.

- They would act as 'force multipliers', thus expanding the capacity and reach of the armed forces.

- They would be able to reach inaccessible areas and survive in inhospitable environment and could be deployed for hazardous tasks that are too risky for combatants.

- They could be deployed for a longer period of time, and would thus cost less than combatants.

- Their destruction would not result in the loss of irreplaceable human life.

- They would be able to use lethal force more conservatively than combatants.

- They would be more accurate and effective than a combatant, thus reducing collateral damage and other mistakes made by a combatant.

- Unlike combatants, they would not suffer from fear or fatigue, and not have the desire for revenge.

- They could be used for casualty evacuation in hazardous combat situations.[9]

- With advancement in robotics, future weapon systems would be better able to comply with the rules of IHL than combatants.[10]

- Their deployment would increase transparency and accountability in operations.

It has also been claimed that robots would help in boosting human capacities in terms of mobility, load capacity, and perception of the environment. They would enhance freedom of action, thus increasing the efficiency of the armed forces. Advocates of LAWS have cited a few examples where a human operator failed to identify a target, resulting in the killing of innocents or the destruction of civilian objects.[11]

Disadvantages

According to Human Rights Watch, the potential advantages of robotic weapons would be offset by the lack of human control over such weapons.[12] The ability of such weapons to perceive their surroundings, reason and communicate is limited. Thus, they would never be able to select and strike targets on the basis of their ability to analyse a complex situation during an armed conflict. It would be impossible for a robotic weapon system to use the tenets of mercy, identification, and morality as human beings are able to do. How would a machine distinguish between a military doctor who is armed for self-defence and a combatant? How well could a robot adapt to changing circumstances, e.g., when a combatant is rendered *hors de combat*?[13] A large section of robotic engineers, ethical analysts, and legal experts are of the firm belief that robotic weapons will never meet the standards of distinction and proportionality required by the laws of war. Besides, in case of a violation of the laws of war by such a weapon, it would be difficult to fix responsibility.[14]

Perhaps the most powerful objection to robotic weapons is moral and ethical in nature. Christof Heyns, the UN Special Rapporteur on extrajudicial killings, has said, "It is an underlying assumption of most legal, moral and other codes that when the decision to take life or to subject people to other grave consequences is at stake, the decision-making power should be exercised by humans."[15] A number of analysts feel that it is morally wrong to give machines the power to decide who lives and who dies on the battlefield. Another disadvantage of robots is their vulnerability to cyber attacks, signal spoofing and electronic jamming. An adversary may disrupt or even take control of an autonomous weapon and use it for unintended activities. Terrorist organizations and rogue armies may also use autonomous weapons in an unrestrained manner. The opponents of robotic weapons fear that emerging technology might trigger a new arms race and encourage leaders to use force rather than resolve conflicts peacefully. They are of the strong opinion that increased lethal autonomy would encourage conflict by lowering war costs and increasing opportunities for accidental engagement.[16]

Future of Autonomous Weapon Systems?

There has been a growing demand from specialists in the field of artificial intelligence, arms control, and human rights to limit the development and use of robotic weapons.[17] The International Committee for Robot Arms Control (ICRAC) is of the opinion that warfare could be accelerated by automated systems, undermining the capacity of human beings to make responsible decisions. According to the International Committee of the Red Cross (ICRC), the development of a truly autonomous weapon system that can implement IHL represents a monumental programming challenge that may well prove impossible. Further, the deployment of LAWS would represent a paradigm shift in the conduct of hostilities.[18] In 2013 the "Campaign to Stop Killer Robots" was launched for a preemptive ban on LAWS. The campaign is supported by many international NGOs, such as Human Rights Watch, Article 36, and International Committee for Robot Arms Control.

In July 2013, the United Nations Secretary-General's Advisory Board on Disarmament Matters drew attention to the increasing trend towards automation of warfare and the development of LAWS.[19] The Advisory Board has raised the following concerns about autonomous weapons:

> The increasing trend towards automation of warfare and the development of lethal autonomous weapon systems gave rise to a wide range of legal, ethical or societal concerns that had to be addressed. ....
> the ability of such weapons to conform to existing law (including IHL, human rights law or general international law); potential problems

associated with the design of future fully autonomous weapons that could require disarmament action, or the ethical limits to robotic autonomy in deciding on the life or death of a human.

The Meeting of High Contracting Parties[20] on Certain Conventional Weapons (CCW) held in Geneva in 2014 agreed on a new mandate on LAWS. It was followed by an informal meeting of experts in April 2015 wherein the questions related to emerging technologies in the area of lethal robots were discussed in the context of the objectives and purposes of the CCW. The Convention currently includes five protocols[21] and demands are now being voiced for a sixth protocol banning the use of lethal robots. The matter was also raised in the UN Human Rights Council (UNHRC) as this body deals in particular with the implications of robotic weapons for human rights. However, in case no protocol on banning such weapons should be agreed upon within the CCW, as appears likely, civil-society groups have suggested that a convention on banning their use could be developed outside this framework. Examples of such conventions include the Ottawa Treaty, the Anti-Personnel Mine Ban Convention and the Convention on Cluster Munitions. There is a possibility that the international community reaches an agreement on a treaty limiting or banning robotic weapons.

One of the main opponents of such a treaty is the US, which believes that LAWS are a useful tool for strengthening national security. The UK and Israel, the other leading developers of robotic weapons, too are of the view that a treaty is unnecessary because they already have internal weapons review processes that ensure compliance with international law.

In December 2007 the US Department of Defence published the Unmanned Systems Roadmap 2007-2032, charting out plans for the development of robotic systems for the next 25years.[22] The US Air Force publication, Autonomous Horizons: System Autonomy in the Air Force – A Path to the Future states:[23]

> Autonomous systems provide a considerable opportunity to enhance future Air Force operations by potentially reducing unnecessary manning costs, increasing the range of operations, enhancing capabilities, providing new approaches to air power, reducing the time required for critical operations, and providing increased levels of operational reliability, persistence and resilience.

In general, military lawyers are of the opinion that it is pointless to debate on robotic weapons at this juncture as the weapon systems have not yet been fully developed. According to them, it is possible that future robots will be able to identify targets better than human beings, respond more rapidly and accurately and cause less collateral damage. However, if such weapons are allowed to develop, it would be meaningless to discuss their acceptability in the future as it would be too late to impose effective restrictions on them by then. Further, no state would be willing to sign a treaty prohibiting or banning their use after spending a huge amount on the development of such weapon systems.

According to reports, some 40 countries are in the process of developing robotic weapons, though only a few have declared their intention openly. The most important ethical question that arises with regard to lethal robotic weapons is whether the decision over life and death can be left to an automated system. Other related questions are: Would such a weapon system be legal under the domain of international humanitarian law? Would the recently adopted Arms Trade Treaty be able to regulate illicit trade in such weapons? Would the deployment of such weapons threaten human rights? Who would be responsible for any serious violation by a robotic weapon system? Do States have an obligation under international law to review the legality of a new weapon or a method of warfare before its deployment? These are the questions that are debated in this book.

Several terms have been used to label future robotic weapons, for example, killer robots, fully autonomous weapons, fully autonomous weapons systems, unmanned lethal weapons, lethal autonomous robots and lethal autonomous weapons systems. The term **Lethal Autonomous Weapon Systems (LAWS)** has been used in this book.

Outline of the Book

Though autonomous military systems are being developed in a number of countries, many terms used in robotic technology remain undefined. **Chapter II** contains a discussion on various terminologies related to autonomous weapon systems. Weapon systems with significant autonomy in the critical functions of selecting and attacking targets are already in use. Besides, the armed forces of at least 40 countries are investing heavily in research and development related to LAWS. **Chapter III** provides a brief account of some robotic weapon systems that are in use or under development.

In today's armed conflicts, as in Syria, it is not easy to distinguish between combatants and civilians mainly because of the absence of clear battle lines and the blending of non-state actors with the civilian population. **Chapter IV** explores the possibility of LAWS being able to comply with the basic principles of IHL, i.e., distinction, proportionality, and military necessity. Besides analyzing certain ethical issues related to the use of LAWS, this chapter also discusses the accountability of a military commander since such weapons would be, by definition, out of the control of their operators.

The binding customary rules of the laws of war prohibit the use of weapons which are incapable of distinguishing between combatants and civilians and are likely to cause superfluous injury or unnecessary suffering. It is in the interest of states to carry out legal reviews of proposed new weapons to ensure that their armed forces would be capable of conducting hostilities in accordance with international legal obligations. The need for a legal review of LAWS has been discussed in **Chapter V**, while international concerns over the development and regulation of future LAWS has been analysed in **Chapter VI. Chapter VII,** the concluding chapter, makes certain recommendation for the UN, the States and the armed forces on the future use of autonomous weapon systems.

The book also contains two appendices. **Appendix A** lists the provisions of international humanitarian law that are related to the use of LAWS in armed conflict. Taking into consideration the many concerns about LAWS, the author feels a pre-emptive ban would eliminate the need to deal with the foreseeable problems. Protocol VI to the CCW, banning the development of LAWS, has been drafted by the author for the consideration of the review conference and is placed at **Appendix B**.

Endnotes

1   The term 'Sepoy' has its origin in Urdu and Persian word "sipahi" meaning soldier. It is used in the militaries of Nepal, India, Pakistan and Bangladesh for a soldier of the army.

2   The word robot comes from the Slavic word 'robota', which meaning labour. The word robotics was introduced to the public by Czech writer Karel Capek in his play RUR (Rossum's Universal Robots), which was published in 1920.

3   Rus Daniela, The Robots Are Coming: How Technological Breakthrough Will Transform Everyday Life, *Foreign Affairs*, July/August 2015, p. 2.

4   Robotic vehicles cannot yet handle all the complexities of driving, for instance, they cannot drive in   inclement weather and complex traffic situations. These issues are the focus of ongoing research. Rus Daniela, The Robots Are Coming: How Technological Breakthrough Will Transform Everyday Life, *Foreign Affairs*, July/August 2015, p. 3.

5   Industrial robots have been in use for about 50 years. The first industrial robot was used for material handling in a General Motors facility. Some applications of industrial robots are material handling; spot welding and arc welding; assembly operations; dispensing functions like painting, gluing, applying adhesive sealing, spraying, etc; and processing.

6   Krishnan Armin. 2009. *Killer Robots: Legality and Ethicality of Autonomous Weapons*, USA: Ashgate.

7   Arkin Ronald C. 2009. *Governing Lethal Behaviour in Autonomous Robots*, USA: Chapman & Hall/CRC.

8   Ozanne Eric, "Robotization of the Combined Arms Task Force: Perspectives", in Doare Ronan, Jean-Paul Hanon and Gerard de Boisboissel (ed). 2014. *Robots on the Battlefield: Contemporary Perspectives and Implications for the Future*, Combat Studies Institute Press, US Army Combined Arms Centre, Kansas, p. 234-235.

9   A report of the US Army's Training and Doctrine Command (TRADOC) titled "Force Operating Capabilities," states: "Future soldiers will utilize unmanned vehicles, robotics, and advanced standoff equipment to recover wounded soldiers from high-risk areas, with minimal exposure." The US Army Training and Doctrine Command (TRADOC) Pamphlet No 525-66, Force Operating Capabilities, 7 March 2008.

10   According to Arkin, autonomous weapon systems could eventually comply with IHL better than human soldiers and such weapon systems are possible in the next two to three decades. For more details see: Arkin Ronald C. 2009,

Governing Lethal Behaviour in Autonomous Robots, USA: Chapman & Hall/ CRC.

11 For example, in the shooting down of Iran Air Flight 655 by the Navy ship USS Vincennes in 1988, the computer system Aegis, which was used, could be set to a number of different modes, which had varying degrees of automation. However, in every mode, human operators had the ability to override the computer. On 3 July 1988, Aegis' radars detected Iran Air Flight 655. The course, speed, radar broadcast, and radio signal coming from the plane indicated that it was a commercial civilian flight. Aegis, which was in semi-automatic mode, registered 655 as an F-14 Fighter, a plane half the size. While all of the evidence should have made it clear that 655 was a civilian plane, not one of the people on board the USS Vincennes were willing to challenge the computer's wisdom. They trusted it and authorized it to shoot. Afterwards, the crew realized that they had shot down a civilian plane, killing all 290 passengers including 66 children. Grut Chantal, The Challenge of Autonomous Lethal Robotics to International Humanitarian Law, *Journal of Conflict & Security Law*, Vol. 18, No. 1, 2013, p. 5–23.

12 Q&A on Fully Autonomous Weapons, Human Rights Watch, 21 October 2013, available at: https://www.hrw.org/news/2013/10/21/qa-fully-autonomous-weapons#2, accessed 12 June 2014.

13 'Hors de combat' is a French term literally meaning 'outside the fight'. The 1977 Additional Protocol I to the Geneva Conventions states that a person is *hors de combat* if he (a) is in the power of an adverse Party; (b) clearly expresses an intention to surrender; or (c) has been rendered unconscious or is otherwise incapacitated by wounds or sickness, and therefore is incapable of defending himself; provided that in any of these cases he abstains from any hostile act and does not attempt to escape.

14 When an autonomous weapon system kills an innocent person unlawfully, who shall we try: the weapon system, its designer, its programmer, the soldier who deployed it, or the military commander who decided that it could be used? A commander is legally responsible for subordinates' actions only if he fails to prevent or punish a foreseeable war crime. Since robots would be, by definition, out of the control of their operators, it is hard to visualize how the commander could be held responsible. *Losing Humanity: The Case Against Killer Robots*, Human Rights Watch, November 2012.

15 Heyns Christof, Report of the Special Rapporteur on extrajudicial, summary or arbitrary executions, UNGA Doc A/HRC/23/47 dated 9 April 2013.

16 'Killer Robots' to Be Debated at UN, BBC, 9 May 2014, available at: http://www.bbc.com/news/technology-27343076, accessed 11 May 2015.

17 In April 2013, an international coalition of nongovernmental organizations (NGOs) launched the Campaign to Stop Killer Robots, calling for a preemptive ban on the weapons. The Campaign, coordinated by Human Rights Watch, now consists of about 50 NGOs in about two dozen countries. It is modelled on the campaigns that led to international bans on antipersonnel landmines, cluster munitions, and blinding lasers.

18 International Humanitarian Law and the challenges of contemporary armed conflicts: Report, 31st International Conference of the Red Cross and red Crescent, Geneva, Switzerland, 28 November-1 December 2011, p. 40.

19 UN Secretary-General, 'Work of the Advisory Board on Disarmament Matters', Report of the Secretary-General, A/68/206 dated 26 July 2013, para 42.

20 The 1980 Convention on Prohibition on the Use of Certain Conventional Weapons Which May Be Deemed to be extremely Injurious or to Have Indiscriminate Effects is usually referred to as the Convention on Certain Conventional Weapons (CCW). It is also known as the Inhumane Weapons Convention.

21 The CCW and its four Protocols prohibit or restrict the use of (i) conventional weapons whose primary effect is to injure by fragments not detectable in the human body by X-rays, (ii) landmines, booby-traps and other devices, (iii) incendiary weapons, and (iv) laser weapons whose combat function causes permanent blindness. The fifth Protocol relates to the clearance, removal or destruction of explosive remnants of war (ERW).

22 Unmanned Systems Integrated Roadmap: FY 2009-2034, The US Department of Defence, p. 210.

23 Autonomous Horizons: System Autonomy in the Air Force – A Path to the Future, Vol. I, Human Anatomy Teaming, The US Air Force Office of the Chief Scientist, AF/ST TR 15-01, June 2015, p. iv.

II Lethal Autonomous Weapon Systems: Definition

With advancement in technology, weapon systems with varying levels of autonomy and lethality have been integrated into the armed forces of numerous states.[1] Though the level of autonomy in these modern weapon systems is increasing day by day, the term 'autonomy' in weapon systems means different things to different stakeholders.[2] In its simplest form, the term means the ability of a machine to perform a task without human input. Autonomy in machines can be understood as the capacity of a robot, following activation, to operate without any external control in some or all areas of its operation for extended periods of time. Autonomy incorporates systems which have a set of intelligence-based capabilities that allow the weapon to respond to situations that were not programmed or anticipated in the design (i.e. decision-based responses). Autonomy represents a significant extension of automation, in which mission-oriented commands would be successfully executed under the circumstances not anticipated, as expected from intelligent human beings when given independence and authority to execute a task. Autonomy can be considered as well-designed and highly capable automation.[3]

Levels of Autonomy

A ten-level spectrum of autonomy in machines has been suggested,[4] the highest level being that of full automation. In this model, at Level 1, a machine is automated and a human being takes all the decisions or actions. At Level 10, a computer decides everything and acts autonomously, ignoring the human operator completely. The machine could be called fully autonomous. At any other level, the system is somewhere in between.

At levels 2 through 4, there is an allocation of decision-making authority between the human being and the machine. At levels 5 through 9, the initial decision-making authority is granted to the machine and the human operator exercises varying levels of approval or veto authority.

Table 1 Levels of Autonomy

Level of Automation	Description
1 Low	The computer offers no assistance; the human operator must take all decisions/actions.
2	The computer offers a complete set of decisions/actions, or
3	Narrows the selection down to a few, or
4	Suggests one, and
5	Executes that suggestion if the human operator approves, or
6	Allows the human operator a restricted time to veto before automatic execution, or
7	Executes automatically, then necessarily informs the human operator, and
8	Informs the human operator after execution only if the latter asks, or
9	Informs the human operator after execution if it decides to do so.
10 High	The computer decides everything and acts autonomously, ignoring the human operator completely.

Depending upon the level of autonomy, Crootof (2015) has divided weapons into four categories: inert weapon systems, automated weapon systems, semi-autonomous weapon systems and autonomous weapon systems. [5]

Table 2 Levels of Autonomy (Crootof)

Level of Autonomy	Weapon Category
1	Inert: rock, sword, pistol
2	Automated: traditional landmine, tripwire sentry gun
3	Semi-autonomous: remotely operated drone, PGMs with autonomous functions to engage specific pre-selected targets
4	Autonomous: C-RAM, Iron Dome, CIWS, missiles which can select targets

Under this classification, inert weapons are objects that require operation by a human being to be lethal. At the second level of autonomy are 'reactive' automated weapon systems. These may be deployed long before they engage a target; they follow pre-programmed instructions, without employing gathered information to draw independent conclusions about how to react. Semi-autonomous weapon systems, the third category, have some autonomous capabilities which may include functions relevant to target selection and engagement, but they cannot independently both select and engage targets. A human operator is required to take some affirmative action to select a specific target for engagement. Weapons in the fourth category, 'autonomous weapon systems', are capable of selecting and engaging targets on the basis of conclusions derived from gathered information and in accordance with the constraints imposed by pre-programming. They do not require any contemporary decisional support by a human operator.[6]

According to the US Air Force, on the basis of the different levels of autonomy, weapon systems could be grouped under the following categories.[7]

- Fully manual: All aspects of task performance are completed by the human being,

- Implementation aiding: These are systems such as flight management systems or smart weapon systems that carry out tasks for the human operator, with the human being making all the decisions.

- Situation awareness support: In this, disparate data are fused to provide integrated information relevant to operator's decisions and goal.

- Decision aiding: The system provides a list of potential options and rates or ranks those options in terms of their suitability, such as with a recommended target list or course of action assessment, (it may or may not select the best option).

- Supervisory control: The system controls all aspects of a function automatically, including taking in information, deciding on correct actions and carrying out those actions, but the human operator can set goals and intervene as and when needed (also called on-the-loop control).

- Full autonomy: This provides for full control over all aspects of a function. There is no scope for human guidance or intervention. For example, the Automatic Ground Collision Avoidance System, currently fielded on F-16s, continuously monitors for impending ground impacts, projects potential escape trajectories, takes control of and executes recoveries at the last possible instant, then returns wings-level control to the pilot.

Keeping in view the above classifications, an 'autonomous system' is a machine, whether hardware or software, which, once activated, performs some task or function on its own.[8] There is no unanimity on the definition of the term 'autonomy' among military lawyers, scientists, weapon developers, human rights activists, philosophers and governments and, therefore, we have no internationally agreed upon definition of the term autonomous weapons.

A few authors have used the term 'semi-autonomous' for those weapons which have some autonomous capabilities like target selection but cannot independently select and engage targets. Such systems may identify a target and await an operator's approval before using lethal force. A human operator will be required to take affirmative action to select a specific target for engagement. Lethal drones fall in this category as they cannot engage a target without approval from a human operator. A drone which could select and engage targets post-deployment without human involvement would be autonomous.

Level of Autonomy in Weapon Systems

Depending upon the level of autonomy in the functioning of weapon systems, different terminologies have been adopted for these systems. The terminologies used by Krishnan,[9] the Human Rights Watch[10] and the US Department of Defence (DOD)[11] have been shown in Table 3.

Table 3 Terminologies for Autonomous Weapon Systems

Autonomy Level	Krishnan	Human Rights Watch	US DOD
A	Pre-programmed autonomy	Human-in-the-loop	Semi-autonomous
B	Limited or supervised autonomy	Human-on-the-loop	Human supervised-autonomous
C	Complete autonomy	Human-out-of-the-loop	Fully autonomous

The US DOD defines the terms as follows.

> **Semi-autonomous weapon system**: A weapon system which, once activated, is intended to engage only individual targets or specific target groups that have been selected by a human operator.

> **Human supervised autonomous weapon system**: An autonomous weapon system that is designed to provide human operators with the ability to intervene and terminate engagements, including in the event of a weapon system failure, before unacceptable levels of damage occur.

> **Fully autonomous weapon system**: A weapon system which, once activated, can select and engage targets without further intervention by a human operator. This includes human-supervised autonomous weapon systems which are designed to allow human operators to override operation of the weapon system, but which can select and engage targets without further human input after activation.

The US DOD describes two types of semi-autonomous weapon systems (SAWS). The first are "systems that employ autonomy for engagement-related functions including, but not limited to, acquiring, tracking, and identifying potential targets; cueing potential targets to human operators; prioritizing selected targets; timing of when to fire; or providing terminal guidance to home in on selected targets, provided that human control is retained over the decision to select individual targets and specific target groups for engagement". The second type of SAWS comprises 'fire and forget' or lock-on-after-launch homing munitions that rely on tactics, techniques and procedures (TTPs) to maximize the probability that the only targets within the seeker's acquisition basket when the seeker activates are those individual targets or specific target groups that have been selected by a human operator.

Autonomous weapon systems, as defined by the US, once activated, can select and engage targets without further intervention by a human operator. They include human-supervised autonomous weapon systems which are designed to allow human operators to override operation of the weapon system, but which can select and engage targets without further human input after activation. This definition has a number of advantages. By using the phrase 'once activated', it highlights how human beings are initially responsible for the decision to deploy LAWS. In this definition, the system's autonomy is determined solely by whether it can select and engage targets without further intervention by a human operator.[12] However, the crucial difference between autonomous and semi-autonomous weapon systems in the US' definitions, i.e. human operator for target selection, is vague and creates confusion.[13]

According to Nils (2013), robotic weapons can be divided into three basic categories,[14] depending on the degree of direct control exercised by a human operator: (i) human-controlled (human-in-the-loop) systems, (ii) human-supervised (human-on-the-loop) systems, and (iii) autonomous (human-out-of-the-loop) systems. In human-controlled systems, weapons are remotely controlled by a human operator. While such robots may be able to independently perform selected tasks delegated to them by their operator (e.g. navigation, systems control, target detection and weapons guidance), they cannot attack without the real-time command of their human operator. As for human-supervised systems, weapons can carry out a targeting process independently of human command, but remain under the real-time supervision of a human operator who can override any decision to attack. In autonomous systems, robotic weapons can search for, identify, select and attack targets without real-time control by a human operator. Such weapon systems can be described as 'automated' when their capability to autonomously detect and attack targets is confined to a comparatively restricted, predefined and controlled environment. When they are capable of autonomously performing these tasks in an open and unpredictable environment, they are described as 'fully autonomous'.

The Human Rights Watch has also provided definitions according to the level of human input and supervision in selecting and attacking targets:[15]

Human-in-the-loop weapons: Robots that can select targets and deliver force only under a human command.

Human-on-the-loop weapons: Robots that can select targets and deliver force under the oversight of a human operator who can override the robots' actions.

Human-out-of-the-loop weapons: Robots that are capable of selecting targets and delivering force without any human input or interaction.

In human-in-the-loop weapon systems, a human operator has to decide about the specific targets to be engaged. The person launching the weapon knows about the specific targets to be engaged and makes a conscious decision that those targets should be destroyed. Modern militaries also use autonomy in several other engagement-related functions prior to the release of weapons. Radars and other sensors help acquire, track and identify potential targets and cue them to human operators. Examples of such weapons could be guided munitions, like projectiles, bombs, missiles, torpedoes and other weapons that can actively correct for initial-aiming or subsequent errors by homing in on their targets or aim-points after being fired, released or launched. An example of man-in-the-loop could be the UK's drone, Taranis. Before the Taranis can deploy lethal force, it must receive permission from a remote human operator. According to Sharkey, 'human-in-the-loop' does not clarify the degree of human involvement; it could be as simple as pressing a button that lights up to say that a target has been detected or it could mean exercising full human judgement about the legitimacy of a target before initiating an attack.[16]

The human-on-the-loop weapon systems use autonomy to select and engage targets where a human operator has not decided that those specific targets are to be engaged, but human controllers can monitor the weapon system's performance and intervene to halt its operation, if necessary. The currently deployed human-supervised autonomous weapon systems include the US Air and missile defence systems, such as the US ship-based Aegis and the land-based Patriot systems. Nearly 40 states employ at least one human-on-the-loop weapon systems.

Human-out-of-the-loop weapon systems use autonomy to select and engage targets where a human operator has not decided that those specific targets are to be engaged, and human controllers cannot monitor the weapon system's performance and intervene to halt its operation, if necessary. According to Human Rights Watch, both human-out-of-the-loop and human-on-the-loop weapons are autonomous weapons.[17] There are a limited number of existing weapons that have a human fully out of the loop for selecting specific targets to be engaged, such that the weapon itself is selecting the targets. These weapon systems use autonomy to engage general classes of targets in a broad geographical area according to pre-programmed rules, and human controllers are not aware of the specific targets being engaged. One can put the Korean SGR-A1[18] or the Israeli Iron Dome[19] in this category.

According to Krishnan (2009), the first level of autonomy, i.e. 'pre-programmed autonomy', indicates that the machine executes a specific function that has been pre-programmed into the system of the machine. Weapons with pre-programmed autonomy have no or very limited capacities to diverge from the pre-set instructions and consequently, operate within very narrow parameters. Examples of machines with pre-programmed autonomy are robots designed for clearing mines, robots tasked with bomb disposal and cave clearance, and systems used for surveillance. In systems with the second level of autonomy, i.e. 'limited or supervised autonomy', the machine operates almost entirely on its own. This means that the variation in its behaviour is greater than in the pre-programmed systems. It can function without continuous human intervention. These systems require human intervention when it comes to the more complex functions, such as targeting, but sometimes also triggering of the weapon. The human operator would function as supervisor as these systems are less capable of dealing with unforeseen situations and circumstances, and where the machine is incapable of proceeding on its own. Machines with 'complete autonomy' can operate completely by themselves, without any human intervention for seeking or attacking targets. They are capable of learning and adapting their behaviour on the basis of previous experiences, and in that sense, built upon an artificial intelligence designed to resemble human intelligence and thought capacity. Complete autonomy means that the operator programmes the machine only with the objective of the mission, and the machine itself will find a solution to it and address the problems which arise on the mission.

However, there could be five levels of human supervisory control of weapons. These are as follows: (i) the human operator deliberates on a target before initiating any attack; (ii) the programme provides a list of targets and a human operator chooses which to attack; (iii) the programme selects a target and a human operator must approve before it is attacked; (iv) the programme selects a target and a human operator has restricted time to veto the choice; and (v) the programme selects a target and initiates an attack without human involvement. According to Sharkey, further research is needed to ensure that human supervisory interfaces make provisions to get the best level of human reasoning needed to comply with the rules of international humanitarian law in all circumstances. [20]

Scharre and Horowith have also used similar terminology for different kinds of autonomous weapons, but have defined them differently.[21]

Human "in the loop" for selecting and engaging specific targets – Weapon systems that use autonomy to engage individual targets or specific groups of targets that a human being has decided are to be engaged.

Human "on the loop" for selecting and engaging specific targets – Weapon systems that use autonomy to select and engage targets where a human being has not decided that those specific targets are to be engaged, but human controllers can monitor the weapon system's performance and intervene to halt its operation, if necessary.

Human "out of the loop" for selecting and engaging specific targets – Weapon systems that use autonomy to select and engage targets where a human being has not decided that those specific targets are to be engaged, and human controllers cannot monitor the weapon system's performance and intervene to halt its operation, if necessary.

Scharre and Horowith have refined these definitions to make a sharper distinction between autonomous and semi-autonomous weapons.

A semi-autonomous weapon – A weapon system that incorporates autonomy into one or more targeting functions and, once activated, is intended to engage only individual targets or specific groups of targets that a human being has decided are to be engaged.

A human-supervised autonomous weapon system – A weapon system with the characteristics of an autonomous weapon system, but with the ability to allow human operators to monitor its performance and intervene to halt its operation, if necessary.

An autonomous weapon system – A weapon system which, once activated, is intended to select and engage targets, where a human being has not decided that those specific targets are to be engaged.

According to Scharre and Horowitz (2015), the idea of a human decision is embedded within each of the above definitions. Further, the definition of an autonomous weapon system is intended to clearly delineate that autonomous weapon systems select and engage targets where a human being has not decided that those specific targets are to be engaged. In the case of an autonomous weapon, the human being decides to launch a weapon to seek out and destroy a general class of targets over a wide area, but does not make a decision about the specific targets to be engaged.[22]

Christof Heyns, the UN Special Rapporteur on extrajudicial, summary or arbitrary executions, considered the technology and defined lethal autonomous robots (LAR) as being "weapon systems that, once activated, can select and engage targets without further intervention by a human operator. The important element is that the robot has an autonomous 'choice' regarding selection of a target and the use of lethal force". According to Christof Heyns, "The measure of autonomy that processors give to robots should be seen as a continuum with significant human involvement on one side, as with UCAVs (Unmanned Combat Air Vehicles), where there is 'a human in the loop', and full autonomy on the other, as with LARs, where human beings are 'out of the loop'." The term 'autonomous' differs from 'automatic' or 'automated'. There are a number of automatic household appliances, like washing machines and cleaners, which operate within a structured and predictable environment. Autonomous systems can function in an open environment, under unstructured and dynamic circumstances, and their actions would be unpredictable in situations of armed conflict.[23]

In its directive, the British Ministry of Defence has defined autonomous systems as follows.

An autonomous system is capable of understanding higher level intent and direction. From this understanding and its perception of its environment, such a system is able to take appropriate action to bring about a desired state. It is capable of deciding a course of action, from a number of alternatives, without depending on human oversight and control, although these may still be present. Although the overall activity of an autonomous unmanned aircraft will be predictable, individual actions may not be.

The directive continues: "Autonomous systems will, in effect, be self-aware and their response to inputs indistinguishable from, or even superior to, that of a manned aircraft. As such, they must be capable of achieving the same level of situational understanding as a human....As computing and sensor capability increases, it is likely that many systems, using very complex sets of control rules, will appear and be described as autonomous systems, but as long as it can be shown that the system logically follows a set of rules or instructions and is not capable of human levels of situational understanding, then they should only be considered to be automated."[24]

Sharkey has defined a lethal fully autonomous robot as one that operates in an open and unstructured environment; receives information from sensors; and processes the information in order to move, select targets and fire – all without human supervision.[25] According to Sharkey, the term autonomy can be very confusing for those not working in robotics. An 'automatic' robot is one which carries out a pre-programmed sequence of operations or moves in a structured environment. An example is a robot arm that is painting a car. In contrast, an 'autonomous' robot is similar to an automatic machine, except that it operates in open or unstructured environments. Asaro (2012) has defined an autonomous weapon system as "an automated system that can initiate lethal force without the specific, conscious and deliberate decision of a human operator, controller, or supervisor".[26] This definition, however, fails to distinguish between weapon systems with different levels of human involvement in the decision to use lethal force. For example, both the Korean SGR-A1 and an antipersonnel landmine could use lethal force without any deliberate decision by a human operator.

The International Committee of the Red Cross (ICRC) also defined automated and autonomous weapon systems in its 2011 report on IHL and challenges in contemporary armed conflicts.[27] The definitions are as follows.

> **Automated weapon systems**: An automated weapon or weapon system is one that is able to function in a self-contained and independent manner although its employment may initially be deployed or directed by a human operator. Although deployed by humans, such systems will independently verify or detect a particular type of target object and then fire or detonate.

> **Autonomous weapon systems**: An autonomous weapon system is one that can learn or adapt its functioning in response to changing circumstances in the environment in which it is deployed.

The ICRC's use of the term 'autonomous weapon' is close to that used by Christof Hynes. According to the ICRC, examples of automated weapon systems include automated sentry guns, sensor-fused munitions and certain anti-vehicle landmines. The capacity to discriminate, as required by IHL, will depend entirely on the quality and variety of sensors and programming employed within the system. The ICRC reserves the term 'fully autonomous weapons' for a class of weapons that does not yet exist. A truly autonomous system would have artificial intelligence that would have to be capable of implementing IHL. The deployment of such systems would reflect a paradigm shift in the conduct of hostilities. Also, a range of fundamental legal, ethical and societal issues would need to be considered before the development or deployment of such systems.

During its expert meeting on "Autonomous Weapon Systems" in 2014, the ICRC maintained that there was no agreed definition of an autonomous weapon system, although various definitions proposed by different countries and organizations shared similar themes. In its proceedings, the ICRC quoted the definitions advanced by the US Department of Defence, the UN Special Rapporteur's report to the Human Rights Council on autonomous weapon systems, and the Human Rights Watch. These have already been discussed.[28]

According to the ICRC, common to all the above definitions is the inclusion of weapon systems that can independently select and attack targets, with or without human oversight. This includes both weapon systems that can adapt to changing circumstances and select their targets and weapon systems that have pre-defined constraints on their operation and potential targets. However, the distinction between 'autonomous' and 'automated' appears to lie only in the degree of freedom with which the weapon system can select and attack different targets. Also common to all these definitions is the exclusion of weapon systems that select and attack targets only under remote control by a human operator. This would exclude the drones being used presently by the US, since targeting and firing is carried out remotely by a human operator. However, the existing remote controlled weapon systems could be developed into fully autonomous systems. In autonomous weapon systems, the weapons can independently select and attack targets, i.e. with autonomy in the 'critical functions' of acquiring, tracking, selecting and attacking targets.[29] A weapon with autonomy in its 'critical functions' means one that can select (i.e. search for or detect, identify, track) and attack (i.e. intercept, use force against, neutralize, damage or destroy) targets without human intervention.[30]

Keeping in view the ambiguities and problems of defining autonomous weapon systems, Crootof (2015) has suggested a new definition from the perspective of the law of armed conflict. According to her, "An 'autonomous weapon system' is a weapon system that, based on conclusions derived from gathered information and pre-programmed constraints, is capable of independently selecting and engaging targets." The term 'weapon system' in this definition denotes a combination of one or more weapons with all the related equipment, materials, services, personnel, and means of delivery and deployment required for self-sufficiency. The clause "based on conclusions derived from gathered information and pre-programmed constraints" attempts to distinguish between 'automated' and 'autonomous' weapon systems. This clause highlights the fact that an autonomous weapon system, like any robotic or code-based system, is controlled by a programme. The clause "is capable of" clarifies that the system "is capable of selecting and engaging targets without further intervention by a human operator".[31]

These definitions accord different levels of autonomy to various kinds of autonomous systems; however, some set the bar for autonomy too low, blurring important distinctions between different levels of human involvement. Others set the bar too high, effectively defining autonomous weapon systems out of existence and ignoring various issues associated with the weapon systems in use today. Some definitions propose distinctions which are based on engineering or philosophy, but which are less appropriate when attempting to evaluate relative levels of human control over the use of lethal force. The above definitions refer to "autonomous weapon system" as a category of weapons that are capable of operating and launching attacks without human intervention or guidance.

According to the author, **lethal autonomous weapon systems are those which select and engage targets without a human operator, wherein lethal force is directed at human beings.** This definition, as compared to the others, appears more appropriate and has been used for discussion in the book. Such weapons do not exist yet; however, the ongoing research in robotics and technology suggests that the concept of LAWS is not far away. Although the debate on LAWS is in its early stages, its opponents and proponents have made diverging claims on the legality, morality and ethics of such weapons when used in an armed conflict.

The next chapter discusses certain autonomous weapon systems. It provides details of a few such systems, which are under development or are being used by the armed forces. Information on them is in the public domain.

Endnotes

1   Crootof Rebecca, The Killer Robots are Here: Legal and Policy Implications, *Cardozo Law Review*, Vol. 36, 2015, p. 1837-1915.

2   'Autonomy' has vastly different meanings in different fields. A political scientist might define autonomy as the ability to be self-governing; a philosopher might focus on an entity's moral independence; and an engineer might be concerned with a machine's level of dependence on human beings in completing different tasks. Due in part to these differing understandings

of autonomy, various stakeholders in the debate over banning autonomous weapon systems often speak past each other. Crootof Rebecca, The Killer Robots are Here: Legal and Policy Implications, *Cardozo Law Review*, Vol. 36, 2015, p. 1844.

3   Autonomous Horizons: System Autonomy in the Air Force – A Path to the Future, Vol. I, Human Anatomy Teaming, The US Air Force Office of the Chief Scientist, AF/ST TR 15-01, June 2015, p. 4.

4   Parasuraman Raja, Sheridan Thomas B. and Christopher D. Wickens, A Model for Types and Levels of Human Interaction with Automation, IEEE Transactions on Systems, Man, and Cybernetics-Part A: Systems and Humans, Vol. 30, No. 2, May 2000, pp. 286–296.

5   Crootof Rebecca, The Killer Robots are Here: Legal and Policy Implications, *Cardozo Law Review*, Vol. 36, 2015, p. 1864.

6   Crootof Rebecca, The Killer Robots are Here: Legal and Policy Implications, *Cardozo Law Review*, Vol. 36, 2015, p. 1864–1865.

7   Autonomous Horizons: System Autonomy in the Air Force – A Path to the Future, Vol. I, Human Anatomy Teaming, The US Air Force Office of the Chief Scientist, AF/ST TR 15-01, June 2015, p. 11.

8   Scharre Paul and Horowitz Michael C., An Introduction to Autonomy in Weapon Systems, Working Paper, The Centre for a New American Security, February 2015, p. 5.

9   Krishnan Armin. 2009. *Killer Robots: Legality and Ethicality of Autonomous Weapons*, USA: Ashgate.

10   *Losing Humanity: The Case Against Killer Robots*, Human Rights Watch, 2012.

11   Autonomy in Weapon Systems, the United States, Department of Defence Directive 3000.09.

12   Crootof Rebecca, The Killer Robots are Here: Legal and Policy Implications, *Cardozo Law Review*, Vol. 36, 2015, p. 1847.

13   According to Gubrud, the operators of autonomous machines accept moral and legal responsibilities to maintain control, and are held accountable for the consequences if they fail to do so. Thus the principle of human control corresponds with the principle of human responsibility. Just as commanders are not responsible for crimes committed by soldiers under their command, unless the commander directly ordered the criminal actions, so human beings cannot accept responsibility for decisions made by machines. But if human beings are to make the decisions, it is their responsibility to maintain control of the machines. Mark Gubrud, Autonomy Without Mystery: Where Do You

Draw the Line? Available at: http://gubrud.net/, accessed 10 November 2015.

14 Melzer Nils, *Human rights implications of the usage of drones and unmanned robots in warfare*, Directorate-General for the External Policies of the Union, European Parliament's Subcommittee on Human Rights, 2013, p. 7.

15 *Losing Humanity: The Case Against Killer Robots*, Human Rights Watch, 2012, p. 2.

16 Sharkey Noel, Towards a principle for the human supervisory control of robot weapons, 2014, available at: https://www.mini-symposium-tokyo.info/ICRA2014/sharkey2014.pdf, accessed 7 July 2015, p. 5.

17 *Losing Humanity: The Case Against Killer Robots*, Human Rights Watch, 2012, p. 2.

18 The South Korean stationary Robot SGR-A1 is equipped with voice and gesture recognition technology. If an enemy approaches the robot, the SGR-A1 can command the person to put his/her hands up and surrender. If the person does not put his/her hands up—as determined by the gesture recognition technology—the robot sends a signal to a human operator, who can choose to use lethal force. Therefore, a human operator makes the final decision and initiates the use of lethal force in man-in-the-loop technology. However, the SGR-A1 can also operate as a man-off-the-loop or fully autonomous weapon system. For more details on the SGR-A1, see Chapter III.

19 For more details on the Israeli Iron Dome, see Chapter III.

20 Sharkey Noel, Towards a principle for the human supervisory control of robot weapons, 2014, available at: https://www.mini-symposium-tokyo.info/ICRA2014/sharkey2014.pdf, accessed 7 July 2015, p. 5.

21 Scharre Paul and Horowitz Michael C., An Introduction to Autonomy in Weapon Systems, Working Paper, The Centre for a New American Security, February 2015, p. 8.

22 Scharre Paul and Horowitz Michael C., An Introduction to Autonomy in Weapon Systems, Working Paper, The Centre for a New American Security, February 2015, p. 16.

23 Heyns Christof, Report of the Special Rapporteur on extrajudicial, summary or arbitrary executions, UNGA Doc A/HRC/23/47 dated 9 April 2013, p. 8.

24 *Unmanned Aircraft Systems: Terminology, Definitions and Classification*, The UK Ministry of Defence, Joint Directive Note 3/10, May 2010.

25 Sharkey Noel, Automating Warfare: Lessons Learned from the Drones, *Journal of Law, Information & Science*, Vol. 21, No. 2, 2012, p. 2.

26   Asrao Peter, On Banning Autonomous Weapon Systems: Human Rights, Automation, and the Dehumanization of Lethal Decision-Making, *International Review of the Red Cross*, Vol. 94, No. 687, (2012), p. 694.

27   International Humanitarian Law and the challenges of contemporary armed conflicts: Report, 31st International Conference of the Red Cross and red Crescent, Geneva, Switzerland, 28 November–1 December 2011, p. 39.

28   *Autonomous weapon systems: Technical, military, legal and humanitarian aspects:* Expert meeting, Geneva, Switzerland, 26–28 March 2014, p. 62–64.

29   There are already some weapon systems in use today that have autonomy in their 'critical functions' of identifying and attacking targets. For example, some defensive weapon systems have autonomous modes to intercept incoming missiles, rockets, artillery shells or aircraft at close range. These weapons can be fixed in place and operate autonomously for short periods of time, in narrow circumstances (e.g. where there are relatively few civilians or civilian objects), and against limited types of targets (i.e. primarily munitions or vehicles). However, in the future, autonomous weapon systems could operate without any limitations, encountering a variety of rapidly changing circumstances and possibly targeting human beings directly. The ICRC has urged states to consider the fundamental legal and ethical issues related to the use of autonomous weapon systems before they are further developed or deployed in armed conflict, as required by IHL. Report of the ICRC Expert Meeting on 'Autonomous weapon systems: technical, military, legal and humanitarian aspects', 26–28 March 2014, Geneva, 9 May 2014.

30   Autonomous weapon systems - Q & A, Geneva: ICRC, 12 November 2014, available at: https://www.icrc.org/en/document/autonomous-weapon-systems-challenge-human-control-over-use-force, accessed 10 June 2015. According to ICRC, the advantage of this broad definition, which encompasses some existing weapon systems, is that it enables real-world consideration of weapons technology to assess what may make certain existing weapon systems acceptable – legally and ethically – and which emerging technology developments may raise concerns under IHL and under the principles of humanity and the dictates of the public conscience.

31   Crootof Rebecca, The Killer Robots are Here: Legal and Policy Implications, *Cardozo Law Review*, Vol. 36, 2015, p. 1854.

III   Autonomous Weapons in Use

Introduction

The armed forces in at least 40 countries are investing heavily in research and development towards testing and deploying automated weapon systems. It is believed that automated weapon systems will play a dominant role in warfare in the future.[1] For instance, the US Army is planning to reduce the number of personnel employed and adopt more robots over the coming years. The number of personnel is expected to shrink from 540,000 to 420,000 by 2019.[2] A few militaries which have developed automated systems are taking the help of robots for risky work. The automated systems used in the armed forces are usually deployed within integrated systems. These robotic systems have different shapes and sizes, according to their purposes, and may be fully autonomous machines or remote-controlled. They may be engaged in different missions, including surveillance, explosive ordnance disposal (EOD), logistics support, search and rescue missions, and combat role. The automated systems integrated in the armed forces could be land-based, air-based, made for operating on the surface of water or underwater, or operated remotely by humans.[3] They require that human operators take decisions on the application of lethal force. However, future robotic systems will most likely be completely automated even in the matter of taking lethal decisions.[4]

Systems on naval vessels have the ability to automatically target incoming missiles, as they confront multiple, inbound, high-speed missiles. Unmanned aircraft systems (UAS) are most often used for intelligence, surveillance and reconnaissance (ISR) missions. Unmanned ground systems (UGS) can enhance joint-service capabilities of the army, such as with ISR, tanks and transport. Navies also employ unmanned surface vehicles (USV), such as for mine countermeasures, ISR and support.

Their unmanned undersea vehicles (UUV) include, for example, missiles and torpedoes. Some of the known robotic systems[5] are discussed in this chapter.

I Ground-based Automated Systems

TALON

TALON is a lightweight, unmanned, tracked military robot designed and built by Foster-Miller, a company owned by QinetiQ North America. The robot was initially developed to protect combatants and detect and destroy explosive threats. Talon is portable (52 kg), and can be easily transported. It is instantly ready for operation. It is the fastest robot today and can keep pace with a running soldier. TALON has a high payload capacity and payload-to-weight ratio, allowing for the incorporation of a broad array of sensor packages. It has outstanding situational awareness and can hold up to four colour cameras, including night vision and thermal and zoom options.

TALON is a versatile robot designed for missions ranging from reconnaissance to weapons delivery. Its large, quick-release cargo bay accommodates a variety of sensor payloads, making TALON a one-robot solution to a variety of mission requirements. Built with all-weather, day/night and amphibious capabilities, it can operate under the most adverse conditions to overcome almost any terrain. It is highly mobile and can climb stairs, negotiate rock piles, overcome concertina wire and plough through snow. It can also be deployed in law enforcement operations and can be reconfigured to conduct a range of missions, including chemical, biological, radiological, nuclear and explosive (CBRNE), explosive ordnance disposal (EOD), rescue, heavy lift, communications, security, reconnaissance, and detection of mines, unexploded ordinance and improvised explosive devices (IEDs). It also supports special weapons and tactics (SWAT) and military police operations.[6] The robot is controlled through a portable or wearable Operator Control Unit (OCU) that provides continuous data and video feedback for precise vehicle positioning.

The TALON robot is used for bomb disposal. It is operated by radio frequency and equipped with four video cameras that enable troops to determine which areas enemy soldiers occupy. In addition, TALON is waterproof up to 100 feet, allowing it to search for explosives off-land.

TALON was also used to locate victims and debris at the World Trade Centre. Talon was initially used to help with military operations in Bosnia in 2000. It was deployed in Afghanistan and Iraq after the war started, assisting with improvised explosive device detection and removal. Talon robots were used in about 25,000 missions in Afghanistan and Iraq. One Talon was blown off the roof of a Humvee in Iraq while the Humvee was crossing a bridge over a river. The TALON flew off the bridge and plunged into the river below. Soldiers later used its operator control unit to drive the robot back out of the river and up onto the bank and could retrieve it.

TALON Operations is in the process of building several distinct 'families' of robots which will be able to perform a variety of tasks, and which will all be operated with one universal control unit. Today, TALON Ground Robotics includes four 'families' that are easily distinguished by size: small, medium, large and extra large. Their names are Dragon Runner (15 to 50 lb), TALON (80 to180 lb), MAARS (300 to 400 lb) and TAGS-CX (5,000 to 6,000 lb). They are all controlled with one new digital control unit.

The Dragon Runner robot was used widely in Iraq. It is designed to be carried in a bag pack for marines and infantry troops. The robots have a rugged design and are equipped with one or more digital cameras, so they can relay images of the operational theatre back to an operational unit. These robots can be tossed around, climb stairs, be dropped from cars, and move in houses and bunkers. In addition, they can move through tunnels with water, scan for snipers, search buildings, screen people for traces of explosives, etc.

Dragon Robot

Guardium

Israel has developed the "Guardium", a remotely operated vehicle. The Guardium autonomous observation and target intercept system was developed by the joint venture company, G-NIUS Autonomous Unmanned Ground Vehicles, established by Israel Aerospace Industries and Elbit Systems. The Guardium has a unique combination of size, payload capacity, reach, dexterity, manoeuvrability, operability and ability to negotiate a wide variety of urban and rural terrain. The Guardium's hybrid extending/folding arm has been designed to be strong and robust, allowing for the deployment of significant payloads with the arm extended and up to 30kg with the arm retracted. The arm extension capability avoids deployment problems in confined spaces, unlike the problems often encountered with unfolding arm systems. The Guardium chassis has been developed to offer a highly stable and manoeuvrable platform. The track modules are able to move continuously through 360 degrees as independent front and rear pairs, ensuring great versatility in positioning, which, in turn, provides obstacle negotiation capability and extremely stable stair climbing. The ability to significantly change the vehicle's footprint ensures stability, allows extended arm reach when required and aids adjustment to the minimum size when accessing confined spaces. Off-road wheels are fitted to enhance cross-country capability, whilst the track modules are retained to assist in obstacle negotiation. The Guardium has a quick-change equipment carrier on the payload head to allow fitting of a variety of equipments, including disruptors, X-ray systems, laser range finder, chemical agents or explosive detectors and ECM equipment. A number of power supplies and data ports are positioned at the base of the arm to allow for the connection of customer-specified equipment.

The system employs autonomous unmanned ground vehicles (UGV) which can be operated from a command centre, carry out routine patrols and quickly respond to evolving emergencies. It can suppress suspicious elements close to the perimeter and hold them back until manned security forces arrive, or use various forceful means to eliminate the threat, if applicable.

The Guardium uses the TomCar chassis. The vehicle is equipped with an automated tactical positioning system and can operate autonomously on- and off-road, at speeds of up to 80 km/h. It can carry a payload of up to 300 kg, including a light armour shield to protect vital systems. The UGV can carry a wide variety of sensors, including video and thermal cameras, with auto-target acquisition and capture, sensitive microphones, powerful loudspeakers and two-way radio. The vehicle can also be equipped with lethal or less than lethal weapons which can be directed and operated from the Main Control Centre (MCC). A fleet of Guardium vehicles can be used as sentries, controlled from the MCC, from where they are launched on routine patrols or ambushes, or to operate in response to news of events received from an early warning or perimeter defence system. The MCC is also provided with automatic tactical area definition, by terrain, doctrine and intelligence, which assist in the preparation of the operational planning and programming for USVs. Each USV can also be manually controlled by remote control.

Guardium

SWORDS

The Special Weapons Observation Reconnaissance Detection System (SWORDS) is an armed version of TALON, carrying a machine gun or a grenade launcher and suited more for defence purposes rather than offensive actions. The SWORDS, designed by Massachusetts-based defence contractor Foster-Miller, was first tried in Iraq in June 2007. It is a ROV (remotely operated vehicle) and not fully autonomous. It is three feet tall and rolls on two tank treads. SWORDS is presently fitted with an M249 machine gun that fires 750 rounds per minute. It can also accommodate other powerful weapons, including a 40-mm grenade launcher and an M202 rocket launcher. Five cameras enable an operator to control SWORDS from up to 800 meters away with a modified laptop and two joysticks. The control unit also has a special "kill button" that turns the robot off should it malfunction. SWORDS is a primitive robot and future models have been tested to develop more sophisticated versions in the future.

iRobot UGVs

Founded in 1990 by Massachusetts Institute of Technology roboticists, iRobot has designed and delivered more than 5,000 defence and security UGVs to researchers, police and armed forces worldwide. Its military models 310 SUGV, 510 Packbot and 710 Kobra are claimed to be force multipliers that perform dangerous search, reconnaissance and bomb-disposal missions. These autonomous systems can inspect and detect explosive/bomb materials inside luggage, suspicious packages, improvised explosive devices (IED) and other containment devices. They contain numerous high-resolution colour camera systems with low-light or thermal vision and support an expanding array of explosive detection sensors that enable real-time detection of trace explosives from down-range.

Modular, adaptable and expandable, the iRobot 510 PackBot can perform bomb disposal, surveillance and reconnaissance, CBRN detection and HazMat handling operations. Quickly configured on the basis of mission needs, PackBot easily climbs stairs and navigates narrow passages with sure-footed efficiency, relaying real-time video, audio and sensor data while the operator stays at a safer, standoff distance. Packbot could be characterized as a smaller, lighter and tougher version of TALON. Like TALON, it can be used to reveal and remove explosives, but it is more suitable for locating enemy soldiers thanks to its endurance and long arm with a camera, which allows it to look through windows or behind corners. Controlled by a Pentium processor that has been designed specially to withstand rough treatment, Packbot's chassis has a GPS system, an electronic compass and built-in temperature sensors. It also uses flippers to get over obstacles and even stairs. Packbot has to be remotely controlled by an operator.

iRobot 510 PackBot

The iRobot 710 Kobra is a powerful, rugged, fast robot that supports or carries multiple and heavy payloads while not sacrificing mobility or operation over rough terrain and stairs. Designed with power, agility and flexibility in mind, the iRobot 710 Kobra can lift in excess of 330 pounds, and negotiate around and through obstacles. It is equipped with numerous payloads to expand the operational area.

BigDog/MULE

BigDog is a mule-sized robot designed to help soldiers carry loads through rough terrain It is a dynamically stable four-legged robot created in 2005 by Boston Dynamics with Foster-Miller, NASA and the Harvard University Concord Field Station. It is 3 feet (0.91 m) long, stands 2.5 feet (0.76 m) tall, and weighs 240 lb (110 kg). It is capable of traversing difficult terrain and running at about 6.5 km per hour. It can carry about 150 kg and climb a 35 degree incline. Its movement is controlled by an onboard computer that receives inputs from the robot's various sensors. Navigation and balance are also managed by the control system.[7]

BigDog uses four legs to get over obstacles and stabilize itself. A stereo vision system allows it to follow a specific person and to create a 3D model of the terrain surrounding it. This enables the robot to identify a safe path forward, and can also enable it to calculate the distances of any gaps or caverns, and whether or not they could be cleared safely with a jump. In laboratory testing, BigDog successfully jumped 1.1 metres with a full payload. BigDog has approximately 50 sensors. These measure the attitude and acceleration of the body, motion and force of the joint actuators, as well as the engine speed, temperature and hydraulic pressure inside the robot's internal engine. Low-level control, such as of the position and force of the joints, and high-level control, such as of the velocity and altitude during locomotion, are both controlled through the onboard computer.

BigDog was funded by the Defence Research Project Agency to serve as a robotic pack mule to accompany soldiers in terrain too rough for conventional vehicles. Instead of wheels or treads, BigDog uses four legs for movement, allowing it to move across surfaces that would defeat wheels. The legs contain a variety of sensors, including joint position and ground contact. BigDog also features a gyroscope and a stereo vision system. Each of BigDog's legs is equipped with four low-friction hydraulic cylinder actuators that power the joints. Built onto the actuators are sensors for joint position and force, and movements were ultimately controlled through an onboard computer which manages the sensors.

Boston Dynamics has designed a refined equivalent of BigDog to exceed it in terms of capabilities and use to dismounted soldiers. In 2012, the militarized legged squad support system (LS3) variant of BigDog, which had the capability of hiking over tough terrain, was also developed. The robotic mule is ruggedized for military use, with the ability to operate in hot, cold, wet, and dirty environments. Further developments were made to make the LS3 more mobile, for example, making it capable of traversing a deep snow-covered hill, or avoiding gunfire and bombs on the battlefield. The objective has been to find an unmanned robotic platform to transport soldiers' equipment and charge batteries for their electronic gear. The vehicle is supposed to carry 1,000 lb (450 kg) of gear, equal to the amount a nine-man infantry squad would need on a 72-hour mission. Attempts are continuing to create more space for equipment, including heavy weapons. BigDog (robotic pack mule) shows great promise in autonomously traversing difficult terrain that must be travelled on foot. This would be of great benefit to today's military persons, who are often required to carry nearly 40 kg of gear—an extremely fatiguing activity in conditions of high temperature and humidity.

The LS3 automatically follows its leader using computer vision, so it does not need a dedicated driver. It also travels to designated locations using terrain sensing and GPS. The US Marine Corps recently demonstrated the performance of the LS3, nicknamed Cujo, which can traverse rocky terrain with its lifelike gallop, and is programmed to follow an operator and detect surrounding objects with its swivelling head of sensors. The Marines displayed Cujo's tricks by using it to conduct resupply missions across terrain difficult to traverse by normal vehicles. Cujo successfully crossed 70%–80% of all terrain traversable by the Marines. A smaller version of bigDog, called SPOT, weighs 160 pounds and was designed to climb up and down hills and stairs. Spot has knees that bend backwards like a mountain goat's. It can carry less weight but is more nimble than bigDog.

The Boston Dynamics' Cheetah robot is the fastest legged robot in the world, surpassing 29 mph, a new land speed record for legged robots. The Cheetah robot has an articulated back that flexes back and forth on each step, increasing its stride and running speed, much like the animal does. The current version of the Cheetah robot runs on a high-speed treadmill in the laboratory, where it is powered by an off-board hydraulic pump and uses a boom-like device to keep it running in the centre of the treadmill.

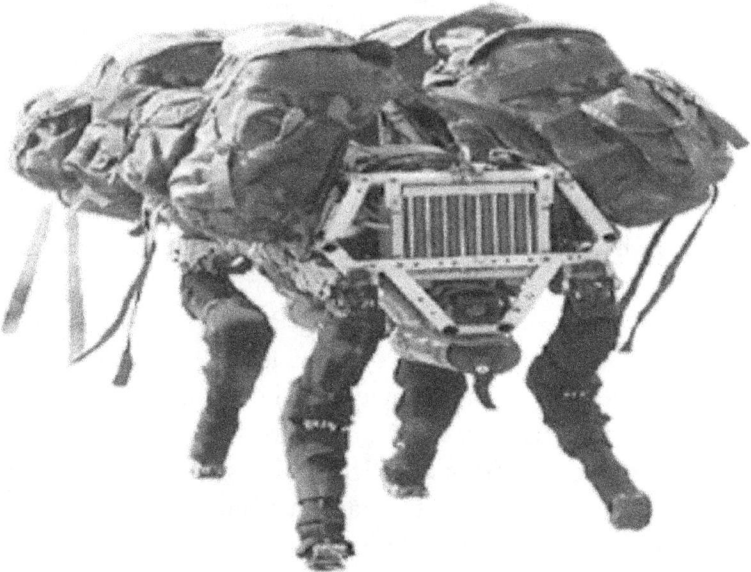

BigDog/Mule

BigDog had a significant ability to keep its balance when travelling on rough and irregular terrain, and when disturbed by outside forces. Although bigDog performed well on rough terrain, it was being modified with larger ranges of limb motion to traverse rougher and steeper terrain with more loads. The main drawback of bigDog was that it was very noisy robot, sounding like a motorcycle. Besides, its load carrying capacity was not enough to fulfill the role of a pack mule. This led the US military to abandon the bigDog project.[8]

MAARS

The Modular Advanced Armed Robotic System (MAARS) is a UGV designed expressly for reconnaissance, surveillance and target acquisition (RSTA) missions to increase the security of personnel manning forward locations. It uses the more powerful M240B medium machine gun and there are significant improvements in its command and control, situational awareness, manoeuvrability, mobility, lethality and safety compared to its SWORDS predecessor. MAARS can be positioned in remote areas where personnel are unable to monitor their security, and can also carry either a direct or indirect fire weapon system.

Agile and combat-ready, MAARS is a technological breakthrough, taking its place on the frontline to keep combatants at a safe distance from enemy fire while effectively executing their security missions. MAARS enables the remote emplacement of RSTA sensors into critical locations up to several kilometers away from the unit, providing early warning, while enabling immediate response if required. Remotely controlled by an operator equipped with a lightweight, wearable control unit, MAARS features multiple onboard day and night cameras, motion detectors, an acoustic microphone, a hostile fire detection system, and a speaker system with a siren to provide optimum situational awareness and alarm.

MAARS' purpose-built system possesses advanced computing power, self-protection features, mobility, modularity, communications systems, sensor payloads, safety features, power management, maintainability and force application capabilities. Advanced processing capabilities and an easy-to-use wearable control system make MAARS simple to operate and powerful. It can even provide multiple options for the escalation of force when required by the rules of engagement (ROE), from non-lethal laser dazzlers and audio deterrents, to less-than-lethal grenades, to lethal fires from the grenade launcher or the medium machine gun. MAARS is also extremely safe and tamper-proof as it can be operated only when receiving properly coded instructions from its operator. MAARS has been developed by QinetiQ North America.

Its operating weight could be up to 350 lb when all sensors, weapons and ammunition are carried. MAARS can be operated from over 800 metres away using either the wearable tactical robotic controller or a laptop controller. Both controllers allow piping of video feed to a TV for observation in a tactical operations centre. Its batteries last 3–12 hours, based on the mission's activities, and it has an integrated sleep mode to save battery power lasting up to a week. It does not have the heat signature of a human being and could 'shoot' intruders with a laser tag gun without being detected itself. The MAARS could be used to launch less lethal weapons, such as tear gas, smoke, star clusters or illuminations. It can also be equipped with a 40-mm grenade launcher, a medium machine gun with 450 rounds of 7.62-mm balls. MAARS can provide remote options to commanders for reconnaissance, assaults, ambushes, hostage rescue, forced entry, booby-trapped areas, detainee riots, site security, improvised explosive device detection and destructions.

MAARS

Crusher

The Crusher is a large-wheeled combat UGV with a payload of 3,600 kg. First introduced in 2006, it is an improved version of the Spinner UGV and was developed to demonstrate technologies for the Future Combat System (FCS) programme's Armed Reconnaissance Vehicle (ARV). Crusher was tested for mobility and autonomy between 2006 and 2008 at various sites around the US (in military bases/posts), where it traversed over 1,400 km. Its payloads include a stabilized, remote-operated small-arms mount with Forward-Looking InfraRed (FLIR), a mast-mounted, stabilized remote surveillance and target acquisition sensor with FLIR, and day cameras. It has a suspended and shock mounted steel skid plate which helps the vehicle to survive encounters with boulders and tree stumps. In front of the vehicle are aids to help it survive collisions with similar other obstacles. Ground clearance is variable and the vehicle can rise up to 60 cm. It is a quiet vehicle since it has a hybrid electric drive system, in which a 60 KW diesel engine charges a lithium ion battery. It boasts of tele-operation and, according to the developer, has full autonomy.

Korean Robotic Sentry: SGR-A1

South Korea has installed a number of machine gun-armed robots to serve as the first line of defence against the potential advance of North Korean soldiers. The stationary robots are operated remotely by humans in a nearby command centre just outside the southern boundary of the DMZ. When the robots' heat or motion detectors sense a possible threat, an alarm goes off in the form of a siren or signal on the screen at the command centre. The operator then uses the robots' video and audio communication equipment to talk with anyone identified by the SGR-A1 before deciding whether to fire the 5.5-mm machine gun. The robots, while having the capability of automatic surveillance, cannot automatically fire at detected foreign objects or figures; a human operator makes the final decision to open fire. The robots can identify targets more than two miles away in daylight, and more than a mile away at night, and can shoot a target as far as two miles away. They are also capable of firing rubber bullets as a warning.

The SGR-A1, developed by Samsung Techwin, makes use of infra-red cameras, alongside a combination of heat and motion sensors. Besides its purpose of detecting intruders, it makes a person think twice about crossing the borders of South Korea illegally. This robot sentry does not distinguish between friends and foes—anyone who crosses the line is considered a potential enemy. Despite the assertion that the robot is under the control of a human operator in a remote location, there exists an automatic modus which enables the robot to decide and open fire on its own. Similar armed robots have been used by the South Korean military on an experimental basis in Afghanistan and Iraq. The whole idea of the SGR-A1 is to let the military robot sentry do the work of its human counterparts in the demilitarized zone between the South Korea and North Korea border, so that there will be a minimal loss of life on the South Korean side, just in case the conflict escalates. The SGR-A1 costs about $200,000.[9]

Israel has also developed Sentry Tech systems along the Gaza border which, in theory, have an autonomous firing mode.

Korean Robotic Sentry: SGR-A1

Daksh

Developed by the Defence Research and Development Organisation (DRDO) of India, Daksh is a battery-operated remote-controlled robot on wheels and its primary role is to recover bombs. It is fully automated. It can climb staircases, negotiate steep slopes, navigate narrow corridors and tow vehicles to reach hazardous materials. Using its robotized arm, it can lift a suspect object and scan it using its portable X-ray device. If the object is a bomb, Daksh can defuse it with its water jet disrupter. It has a shotgun, which can break open locked doors, and it can scan cars for explosives. With a master control station (MCS), it can be remotely controlled over a range of 500 m in line of sight or within buildings. Ninety per cent of the robot's components are indigenous. The Army has placed limited series production orders for 20 Dakshs. In December 2011, the first batch of five units was handed over to General Combat Engineers. The technology has been transferred for production to three firms—Dynalog, Theta Controls and Bharat Electronics Ltd.

Remotely Operated Vehicle - Daksh

There are reports that the DRDO is working to develop robotic soldiers which could be deployed in sensitive border conflict zones, including the so-called line of control dividing India and Pakistan. The robot is likely to have a very high level of intelligence to be able to make a distinction between combatants and civilians. As reported, initially, a human soldier would be in the loop to identify an enemy or a combatant for a robotic soldier, but in due course of time, the robotic soldier would be at the front end and the human soldier would be assisting him. [10]

GRUNT (unmanned GRound UNiTs)

The GRUNT, developed by Frontline Robotics, is a medium-wheeled patrol and surveillance UGV used by Canada and South Korea. A Mobile Autonomous Guard System (MAGS), including two GRUNTs, is based on the Argo All-Terrain-Vehicle, a base station command and control system. The vehicles include COTS sensors and the PC-104 industrial computer. The GRUNT can perform perimeter security checks and surveillance of critical infrastructure. It is an autonomous UGV that has radio communication systems, all-around imaging cameras, night-vision sensors, radar and a continuous navigation system. A Frontline Robotics' Team Intelligence software platform, named Robot Open Control (ROC),

GRUNT (unmanned GRound UNiTs)

enables fully autonomous operation and decision-making collaboration between GRUNTs. The GRUNTs can work together to identify an intruder and disturb and prevent intrusion, and they can observe security threats and communicate with personnel staying in a safe place.

Platform-M

Russia has recently unveiled Platform-M, a 'universal combat platform'. It is a multipurpose weapon system, designed for gathering intelligence, discovering and eliminating stationary and mobile targets, firepower support, patrolling and guarding important sites. Developed by Progress Scientific Research Technological Institute of Izhevsk, this weapon system can be guided, carry out supportive tasks and destroy targets in automatic or semi-automatic control systems. In other words, it can operate in autonomous or semi-autonomous modes. Platform-M is armed with four grenade launchers and Kalashnikov rifles.

Platform-M

II   Unmanned Aerial Systems

Militaries across the world are becoming much more heavily reliant on unmanned aerial systems (UAVs). The UAVs offer an extremely wide range of military applications, such as security and control; aerial reconnaissance; aerial traffic and security watch; battlefield management; telecommunications; all-terrain search and rescue; disaster damage estimation; wide area munition deployment, and for the launch of air-to-ground and air-to-air missiles. Due to the development of unmanned technology in Israel, the country's air force has become a leader in the establishment of unmanned aerial vehicle squadrons. Israel has never confirmed that it has unmanned combat aerial vehicles (UCAVs), and information related to them is classified and protected by the state's military censorship. [11] As known, presently two completely autonomous UAV technologies are being developed. These are the British Taranis and the American X-47 systems.

Taranis

Taranis, named after the Celtic god of thunder, is an unmanned combat aircraft system being developed by the British industry. The project is funded by the UK Ministry of Defence (MoD) and is managed by the MoD's Unmanned Air Systems Project Team in the Defence Equipment and Support organization based in Bristol. First flown in 2013, the unmanned stealth aircraft is 12m long and has a wing span of 10m. Taranis would be capable of delivering weapons to a battlefield in another continent. It will be able to hold an adversary at continuous risk of attack; penetrate deep inside hostile territory; find a target; facilitate either kinetic or non-kinetic influence upon it; assess the effect achieved; and provide intelligence back to commanders.

Taranis demonstrates the ability of an unmanned combat aerial system to fend off hostile attack; deploy weapons deep in enemy territory; and relay intelligence information. Additionally, both the shape and internal technologies help Taranis to remain undetected by enemies. Taranis would reach the search area via a pre-programmed flight path in the form of a three-dimensional corridor in the sky. Intelligence would be relayed to the mission command. When Taranis identifies a target, it would be verified by the mission command. On the authority of the mission command, Taranis would carry out a simulated firing and then return to base via the programmed flight path. At all times, Taranis is under the control of a highly-trained ground crew. The mission commander both verifies targets and authorizes simulated weapons release. Taranis is planned to be operational 'post-2030' and will be used in concert with manned aircraft.

Taranis

X-47B

In early 2013, the US conducted the first ever carrier launch of a low-observable unmanned combat aerial vehicle (UCAV), the 'X-47B', from the USS George H.W. Bush. It was a milestone in aviation history as this was the first carrier-based launch of an unmanned stealth combat aerial vehicle capable of autonomous operation without significant human intervention. The X-47B is a tailless, strike fighter-sized unmanned aircraft developed by Northrop Grumman for the US Navy. It is capable of taking off, flying a pre-programmed mission, and then returning to base in response to its mission operator. The mission operator monitors the X-47B air vehicle's operation, but does not actively 'fly' it via remote control, as in the case of the other unmanned systems currently in operation.

In April 2015, the X-47B once again made aviation history by successfully conducting the first ever autonomous aerial refuelling (AAR) of an unmanned aircraft. AAR unlocks the full potential of what an unmanned surveillance, strike and reconnaissance system can do in support of the Navy. These historical demonstrations solidified the concept of future unmanned aircraft and proved that the X-47B can perform standard missions, like aerial refuelling, and operate seamlessly with manned aircraft as part of the Carrier Air Wing. The X-47B is equipped with an autonomous aerial refuelling system. To date, it has conducted operations aboard three different aircraft carriers: the USS Harry S. Truman (CVN 75), the USS George HW Bush (CVN 77), and the USS Theodore Roosevelt (CVN 71). This revolutionary technology can increase the range and flexibility of future unmanned and manned aircraft platforms. The X-47B has paved the way for the future sea-based unmanned aircraft system by digitizing the carrier controlled environment, achieving precision landing navigation performance, demonstrating a deck handling solution, and refining the concept of operations. It has a wingspan of 62.1 ft and a length 38.2 ft, and flies at an altitude of 40,000 ft. Its range is up to 2,100 nm and its maximum gross takeoff weight is 44,000 lb. The X-47B UCAS is designed to help the Navy explore the future of unmanned carrier aviation. The introduction of carrier-based long-range unmanned stealth combat UCAV reduces the necessity for theatre air bases around the enemy country, such as the bases on which the US campaigns have so far depended to maintain high intense sortie rates. This development will also retain the role and position of aircraft carriers in future US military campaigns.[12]

Mid-air refuelling of an unmanned aircraft by X-47B

Iron Dome

Israel's Iron Dome anti-rocket system has been widely credited with protecting the country's civilian population from projectiles fired by Palestinian militants in the Gaza Strip. During the conflict between Israel and the Hamas and other Gaza-based militants (Operation Protective Edge) in the summer of 2014, the performance of the Iron Dome was praiseworthy. The Iron Dome is a short-range anti-rocket system developed by Israel's Rafael Advanced Defence Systems and originally produced in Israel. Its targeting system and radar are designed to fire its Tamir interceptors only at incoming projectiles that pose threats to the protected areas. The Iron Dome system is not configured to fire on rockets headed towards unprotected areas. Its batteries can be moved to respond to changes in Israeli areas subject to threat.

The Iron Dome is an effective and innovative mobile defence solution for countering short-range rockets and 155-mm artillery shell threats with ranges of up to 70 km in all weather conditions, including low clouds, rain, dust storms or fog. The system uses a unique interceptor with a special warhead that detonates any target in the air within seconds. The Iron Dome radar detects and identifies a rocket or artillery shell and monitors its trajectory. Target data is transmitted to the Battle Management and Weapon Control (BMC) for processing. The threat's trajectory is quickly analysed and the expected impact point is estimated. If the estimated rocket trajectory poses a critical threat, a command is given within seconds and an interceptor is launched against the threat. The interceptor receives trajectory updates from the BMC via uplink communication. The interceptor approaches the target and uses its radar seeker to acquire the target and guides the interceptor within passing distance. The target warhead is detonated over a neutral area, therefore reducing collateral damage to the protected area.

The Iron Dome was declared operational in early 2011. In a week-long Israeli–Hamas conflict in November (Operation Pillar of Cloud), Israeli officials claimed that the Iron Dome intercepted 85 per cent of the rockets fired by Gaza-based militants. Between 2012 and 2014, Israel upgraded the Iron Dome's various tracking and firing mechanisms and expanded the number of batteries deployed from five to nine.

Iron Dome Battery

Because the Iron Dome was developed by Israel alone, Israel initially retained proprietary technology rights to it. The US and Israel have had a decades-long partnership in the development and co-production of other missile defence systems and, as the US began financially supporting Israel's development of the Iron Dome in 2011, it became a partner in its co-production. In March 2014, the US and Israeli governments signed a co-production agreement to enable components of the Iron Dome system to be manufactured in the US, while also providing the US Missile Defence Agency (MDA) with full access to what had been proprietary Iron Dome technology. The US-based Raytheon will be Rafael's US partner in the co-production of the Iron Dome, and Raytheon's facility in Tucson, Arizona may be one of several US sites where production takes place.

The Iron Dome is a cost-effective system that can handle multiple threats simultaneously and efficiently. It has been selected by the Israeli Defence Ministry as the best system offering the most comprehensive defence solution against a wide range of threats in a relatively short development cycle and at low cost.[13]

Brimstone

Brimstone is an advanced rocket-propelled, radar-guided weapon and can seek and destroy armoured targets at long range. Developed for the British military, it has been described as a 'fully autonomous, fire-and-forget' weapon. During the search phase, Brimstone's millimetric wave radar seeker searches for targets, comparing them to a programmed target signature in its memory. The missile automatically rejects returns which do not match this programming and continues searching and comparing until it identifies a valid target or self-destructs. Once launched, the weapon operates autonomously. Brimstone can be programmed to start searching only in target areas, limiting risks to friendly forces. The critical aspects of how human control is exercised over such weapons pertain to the programming of the target parameters and sensor mechanisms, and to the area within which and the time during which the weapon operates independently of human control. The Peace Research Institute Oslo (PRIO) deemed that Brimstone's ability to autonomously select targets was "ill-suited to contemporary operations", especially in Afghanistan, because of the conflict's complex nature; the rules of engagement required that a human being monitor the engagement right up to the impact of the missile.[14] As a result of this challenge, additional mechanisms were put in place to ensure human oversight of final target selection. The linking of meaningful human control to individual attacks is significant because it is in relation to individual attacks that the existing rules of international humanitarian law apply – it is over individual attacks that commanders must make legal judgments.

NBS MANTIS

Germany had deployed the NBS MANTIS Defence Protection System, a short-range, land-based, air defence protection system in forward-operating bases of the German Army in Afghanistan. It is equipped with radar sensors and is capable of independently recognizing flying objects from approximately three kilometers away. The NBS MANTIS instantly and automatically engages the incoming object with guns that are capable of firing 1,000 rounds per minute based on mission-specific, pre-programmed software. It can operate autonomously without human intervention. If necessary, human operators may override it.

Bionic Hornets

It has been reported that Israel is utilizing nanotechnology to develop a 'bionic hornet' (i.e. flying micro robot) capable of targeting, chasing, photographing and killing enemy combatants/insurgents/terrorists. As reported, Israel has decided to further lethal nanotechnology on the basis of its recent experience in fighting guerrillas in Lebanon. The Israelis have apparently come to the realization that it is not economically feasible to employ fighter aircraft to fight Hizbollah fighters armed with rocket ATGMs (anti-tank guided missiles). The bionic hornets can navigate through city streets, alleys and buildings to reach the 'bad guys' and explode, essentially making them like little micro kamikazes.[15] Similarly, the US military SWARM technology would allow one human operator to control a fleet of aircraft that have been developed to respond in a synchronized fashion.

AURA

The AURA is an autonomous unmanned combat air vehicle (UCAV) being developed by the Defence Research and Development Organisation[16] for the Indian Air Force and Indian Navy. The main role of the AURA is to be deployed as an unmanned stealth bomber. The designing of the AURA is to be carried out by the Aeronautical Development Agency (ADA). The UCAV will be capable of releasing missiles, bombs and PGMs. It will act as the ultimate 'force multiplier' and 'game changer' in any battle scenario in the future. The details of the project are not in the public domain.

III Maritime Autonomous Weapon Systems

In the past, the marine environment has been subjected to deep sea oil rigs and underwater cable networks. Historically, the unmanned undersea vehicles (UUVs) of the military were used during the 1950s and 1960s, when the self-propelled underwater research vehicle was used in oceanography.[17] The first practical autonomous unmanned marine system (UMS), used for oceanographic surveys and clearing mines, appeared in the late 1990s. Several technologies have made it possible today to build even more sophisticated autonomous UMSs.[18] In 2003, during Operation Iraqi Freedom, autonomous underwater vehicles were used for the first time in Umm Qasr Harbour for mine warfare operations. Now the states are intruding into the high seas with autonomous technologies in the form of surface and underwater systems known variously as ROVs (remotely operated underwater vehicles), USSVs (unmanned sea surface vessels), UUVs (unmanned underwater vehicles) and AUVs (autonomous underwater vehicles).[19] Most unmanned underwater systems require some level of autonomy due to the fact that standard methods of navigation, such as GPS, do not work underwater. Research is under way to increase autonomous functions for more complex operations.[20]

Technological advancement has permitted the development of autonomous marine weapon systems and the replacement of manned systems by these is a very attractive proposition for the military.[21] Many of the world's conflict flashpoints are on coastal or contested waters. With rising seas, changing weather patterns and other consequences of global warming, access to previously impassable areas will render the maritime environment an increasingly strategic battle space for an ever-growing number of States.[22] Strategic experts are of the view that fully autonomous weapon systems or LAWS are likely to first appear in the maritime environment.[23] However, in order to perform their mission successfully, the LAWS will need to demonstrate several facets of advanced autonomous operation. This includes their compliance with IHL, maritime laws and conventions for safe navigation, and operational reliability.

The sea-based unmanned maritime systems (UMS) can be described according to four dimensions: autonomy, operating mode, size and functions. UMSs can be either free-swimming or tethered to a surface vessel, a submarine, or a larger robot. Tethers simplify the provision of power, control and data transmission, but limit manoeuvrability and range. Recently, developers have built highly autonomous systems that can navigate, manoeuvre, and carry out surprisingly complex tasks. UMSs can operate on the ocean's surface, at or just below the surface, or entirely underwater. Operating above or near the surface simplifies power and control, but compromises stealth.

A wide range of increasingly autonomous military systems is already deployed in the maritime environment, for a variety of missions. These include mine-countermeasures, and intelligence, surveillance and reconnaissance (ISR). These systems enable States to monitor activities of interest and potentially hostile actions over areas that could not be easily covered by human operators. Unmanned surface vehicles can be deployed on missions such as mine-countermeasures, port surveillance, fleet protection and supply delivery. Recent developments in swarming capabilities open up realistic possibilities for automated ship protection and area denial, where autonomous surface vessels operate in defensive postures yet could have offensive capabilities.[24] Increasingly autonomous military systems are also of interest for coastal monitoring, anti-piracy and counter-narcotics operations, as well as wide area searches, such as for Malaysian Airlines flight 370.

The US Navy has devoted particular attention to unmanned underwater vehicles (UUSs) during the past 10–15 years.[25] The Navy has given greater priority to the use of automation to reduce the crew size in warships. As far as their sizes are concerned, the UMSs range from small or "man-portable" (25–50 pounds); to mid-sized (500–3,000 pounds, often designed to operate from existing submarine weapon launch tubes and airlocks); to large (projected in the future to be up to 40 feet long, displacing 10 tons and designed to operate from a pier, off a surface vessel, or from a large weapon tube). UMSs can be equipped with a variety of manipulator arms, capture devices, sensors, communications packages and warheads.

Anti-Submarine Warfare (AWS) Continuous Trail Unmanned Vessel (A-CTUV)

The American Defence Advanced Research Projects Agency programme A-CTUV (Sea Hunter) can detect and track quiet diesel electric submarines for several months at a time. The US Department of Defence Directive 3000.09 permits semi-autonomous weapon systems that do not "autonomously select and engage individual targets or specific target groups that have not been previously selected by an authorized human operator". In order to perform the mission, the Sea Hunter system needs to demonstrate several facets of advanced autonomous operation, including compliance with maritime laws and conventions for safe navigation, autonomous system management for operational reliability, and the rules of IHL.

Anti-Submarine Warfare (AWS) Continuous Trail Unmanned Vessel (A-CTUV)

The A-CTUV is to be in constant contact with other ships and aircraft through a satellite link. If a contact is determined to not be threatening, a sailor can order the vessel to go back on patrol. The craft itself is unarmed, so if an enemy submarine is detected, it will notify other naval assets that can attack and destroy it. If deemed not a threat, the craft can still shadow the submarine, potentially even back to its home port, to deter it from acting aggressively. The A-CTUV is designed to out-endure all diesel-electric submarines, even those equipped with air independent propulsion (AIP). Using large numbers of inexpensive unmanned A-CTUVs is a way of countering submarines as an undersea component of anti-access warfare. In order to comply with the International Regulations for Preventing Collisions at Sea (COLREGS), the A-CTUV has to autonomously identify surface ships at sea.

Large Displacement Unmanned Undersea Vehicle (LDUUV)

The Large Displacement Unmanned Undersea Vehicle (LDUUV) is intended to conduct missions longer than 70 days in the open ocean and littoral seas. It will be a fully autonomous, long-endurance and land-launched vehicle, with advanced sensing for littoral environments. The vehicle's testing is planned in 2018. According to the Navy's ISR Capabilities Division, the LDUUV will reach its initial operating capability as a squadron by 2020 and full-rate production by 2025. The craft's missions will include ISR, acoustic surveillance, anti-submarine warfare, mine counter-measures, and offensive operations.

ONR LDUUV Prototype

The US Navy's Underwater Gliders

These autonomous underwater gliders will not require fuel, but will instead use a process called hydraulic buoyancy, which allows the drone to move up and down and in and out of underwater currents. This will help it move at a speed of about one mile per hour. Carrying a wide variety of sensors, they will be programmed to patrol for weeks at a time, surfacing to transmit their data to the shore while downloading new instructions at regular intervals.

The US Navy plans to use its fleet of deep and shallow water Littoral Battlespace Sensing-Gliders (LBS-G) to acquire critical oceanographic data, which will improve the positioning of fleets during naval manoeuvres. The gliders will also have obvious uses for submarine hunting and hiding, given the effect of temperature layers on sonar propagation.

Littoral Battlespace Sensing-Gliders (LBS-G)

Collectively Behaving Robots

To be of real help in complex military applications, robots should be an integral part of manned systems. They should also be capable of being used massively, in robotic collectives. The tests on Virginia's James River represented the first large-scale military demonstration of a swarm of autonomous boats designed to overwhelm enemies. The boats operated without any direct human control: they acted as a robot boat swarm. This capability points to a future in which the US Navy and other militaries may deploy multiple underwater, surface and flying robotic vehicles to defend themselves or attack a hostile force.

Harvard University scientists have devised a swarm of 1,024 tiny robots that can work together without any guiding central intelligence. Like a mechanical flash mob, these robots can assemble themselves into five-pointed stars, letters of the alphabet and other complex designs. Swarm scientists are inspired by nature's team players—social insects, like bees, ants and termites; schools of fish; and flocks of birds. These creatures collaborate in vast numbers to perform complicated tasks, even though no single individual is actually in charge. These results are believed to be useful for the development of advanced robotic teams or armies.

MK15 Phalanx Close-In System (CIWS)

The US Navy's self-defence system is the MK 15 Phalanx Close-In System (CIWS), which detects missiles and aircraft that have breached a ship's primary defence envelop. The Phalanx system has been used on more than 250 ships since its inception. The Phalanx performs multiple functions autonomously. These include "search, detection, threat evaluation, tracking engagement and kill assessment". It engages incoming anti-ship air missiles without assistance from human operators, although the system does provide for manual override. It serves as the last line of defence if other defence systems fail, and consists of a radar system and a 20-millimetre rapid-fire machine gun.

MK 15

AEGIS System

Another naval system similar to the Phalanx is the AEGIS combat system, which automatically engages targets and destroys incoming missiles. First commissioned in 1983 and now equipping 74 US ships, the AEGIS can simultaneously react to multiple threats, including air, surface and submarine. The Navy also has several other autonomous systems in the pipeline. These include the Long-Range Anti-Ship Missile (LRASM), designed to combat the threat from ballistic missiles that can target US ships from hundreds of kilometres. The LRASM will be capable of independent targeting by relying exclusively on on-board sensors and processing without any prior precision intelligence or GPS. It will be distinctive for its range, ability to precisely target moving objects and avoid enemy defence systems. The Navy has also tested swarm boats that can autonomously surround ships, protect them from incoming enemy attacks, avoid colliding with civilian traffic and attack in a coordinated unit.

The British marine systems are mainly used for mine hunting and disposal. Since 2006, the Royal Navy has used 'Seafox', an unmanned underwater vehicle tethered to a mother ship, for mine hunting and disposal. In 2014, the Navy began testing the 'Hazard', a remote-controlled boat that can deploy a number of unmanned underwater vehicles, including Seafox, to detect and dispose of mines. A number of reports say that the UK has a maritime surveillance capability gap. The British Ministry of Defence is exploring options for providing part of a future capability with an aerial unmanned system. The Royal Navy is developing a maritime UAS.[26]

The Indian Naval Indigenization Plan (INIP) (2015–2030), issued by the Directorate of Indigenization IHQ, MOD (Navy), includes unmanned aerial vehicles (UAV), remotely operated vehicles (ROV) and autonomous underwater vehicles (AUVs). It states that unmanned vehicles will progressively find increasing use in naval applications. The operational spectrum of the UAVs will include reconnaissance, C^2, target discrimination and identification, battle damage assessment, data transfer, electronic countermeasure (ECM), electronic support measure (ESM), electronic counter-countermeasure (ECCM) and combat support /identification in case of shore bombardment and amphibious operations. The UAVs will act as a force multiplier and represent the 'eyes' of naval units in the future, making it possible for them to see in real-time-over-the-horizon. The future capabilities of UUVs would also include the ability to carry a limited range of weapons for attacking detected targets. In the future, surface ships operating in littoral waters can be expected to encounter novel threats, like intelligent sleeping mines, frogman, miniature submarines and intelligent torpedoes. The countermeasures will need to include artificial, remote-controlled 'fish', equipped with explosive loads that can be activated through acoustic means.

The DRDO has successfully developed an autonomous underwater vehicle (AUV), which will aid the Indian Navy in surveying waters and help in the deterrence of hostile ships or submarines. The robotic vehicle is fully pre-programmed—in terms of algorithms and strategy, and mission requirements—and piloted by an on-board computer. There is no control of the vehicle once it is released into water. However, if the AUV deviates from its intended path, the guidance and control systems activate the propellers (thrusters) and control planes to ensure that the vehicle returns to the original trajectory and continue moving along the desired path.

The AUV, which is very small in size and operates around a mother ship from where it is launched, controlled and recovered, is expected to aid the Indian Navy in surveying waters and help in the deterrence of hostile ships or submarines. Developed from a concept vehicle weighing 300 kg, the AUV has two interconnected cylindrical pressure hulls. Its multi-sensor intelligence robotic architecture provides for underwater monitoring and communication. Since the thrusters are inside the pressure hulls, vibration is minimal. With the development of the AUV and such indigenous technology, the DRDO has enabled India to be at par with nations like the US and Japan. The cost of the AUV is roughly $8.4 million and it will augment the underwater surveillance capabilities of the Indian Navy.[27]

DRDO's AUV

Important as these individual technologies are, system integration is likely to be more critical. The biggest challenges for underwater vehicles remain communications and power, along with ocean pressures that stress battery housings and other structures, and deep sea cold and darkness that make any kind of work challenging. It is the combination of sensors, communications, data processing and mechanical design that produces an effective package.

Level of Autonomy in Marine Systems

The US Office of Naval Research (ONR) has defined six levels of vehicle autonomy, summarized by the Committee for the Review of ONR's Uninhabited Combat Air Vehicles Programme. These are:

(i) Fully autonomous: The system requires no human intervention to perform any of the designed activities across all planned ranges of environmental conditions.

(ii) Mixed initiative: Both the human being and the system can initiate behaviours based on sensed data. The system can coordinate its behaviour with the human being's behaviours both explicitly and implicitly. The human being can understand the behaviours of the system in the same way that he or she understands his or her own behaviours. A variety of means is provided to regulate the authority of the system with respect to human operators.

(iii) Human-supervised: The system can perform a wide variety of activities once given top-level permissions or direction by a human being. The system provides sufficient insight into its internal operations and behaviours, allowing it to be understood by its human supervisor and appropriately redirected. The system cannot self-initiate behaviours that are not within the scope of its current directed tasks.

(iv) Human-delegated: The system can perform limited control activity on a delegated basis. This level encompasses automatic flight controls, engine controls, and other low-level automation that must be activated or deactivated by a human being and act in mutual exclusion with human operation.

(v) Human-assisted: The system can perform activities in parallel with human input, thereby augmenting the ability of the human being to perform the desired activities. However, the system has no ability to act without accompanying human input.

(vi) Human-operated: All activity within the system is the direct result of human-initiated control inputs. The system has no autonomous control of its environment, although it may be capable of information-only responses to sensed data.

These modes of operation and the technology required to shift from one level of autonomous operation to another are under development.[28]

Problems with Marine Autonomous Weapon Systems

Maritime warfare demands innovative approaches to achieve strategic leverage. Unmanned and increasingly autonomous lethal systems will permit the deployment of ever greater numbers of maritime objects. This increasing mix of both civilian and military objects and infrastructures is likely to make the marine environment both more complex and cluttered. The operation of increasingly autonomous technologies on the high seas, in territorial or littoral waters, and in Exclusive Economic Zones will need to take into account the various permitted operational activities and the applicable legal regimes.

There are two categories of objects in maritime law relevant to marine autonomy: vessels and weapons. The existing law relevant to armed conflict at sea is primarily built around the concept of a vessel, such as a ship, submarine and landing craft. Vessels are subject to a range of rules and responsibilities under international law. These include the responsibility to: (i) search for and assist wounded, sick and shipwrecked persons after an engagement, (ii) search for persons reported missing by one's opponent, and (iii) respect and provide quarter to an enemy vessel that has signalled its intention to surrender. There are also protections afforded to specific types of vessels. Military vessels are obliged to respect the ships of neutral countries, as well as hospital and scientific ships and those carrying civilians. Under certain conditions, a State may evade the responsibility of avoiding harm to vessels not involved in the conflict by declaring an "exclusion zone" for such vessels.

Technological advancement has made it possible to distinguish between enemy warships and protected vessels on the high seas to a high degree of certainty. Systems operating in the submarine environment can make a distinction by detecting acoustic signals that provide information on the type of ship involved, its speed and direction on the basis of known acoustic signatures for particular types of ships.[29] However, it is not clear whether this high degree of distinction can be maintained in lethal autonomous underwater systems. Further, the lethal autonomous technologies may not be able to fulfil the two major obligations under IHL, i.e., to search for the wounded and sick, and to provide quarter to an enemy vessel that wishes to surrender. As compared to terrestrial and aerial LAWS which are relatively easy to observe via a variety of human and technical means, underwater deployment and operations of lethal autonomous weapons is likely to be less transparent. In addition, the deployment of a highly autonomous system outside communication range and outside an active conflict could lead to accidental collision and increase tension.

Marine LAWS can be used for important missions and can add new capabilities to confound potential adversaries, though they cannot fulfil the needs of the navy completely. Some of the functions for which sea-based autonomous weapon systems could be used are: counter-mine warfare, anti-submarine warfare, information operations, strike warfare and area denial, harbour patrolling to detect intruders, infiltration and payload emplacement, infrastructure inspection and servicing, *ad-hoc* sensor and communication services, environmental monitoring and oceanography for long-term planning and research.[30]

Conclusion

Military systems already have some automated functions, including navigation, take-off and landing, communications and detection. The level of automation a system exhibits can be seen as a spectrum ranging from remotely piloted through to fully autonomous systems. At least 30 nations employ or have in development at least one system of this type. These nations are Australia, Bahrain, Belgium, Canada, Chile, China, Egypt, France, Germany, Greece, India, Israel, Japan, Kuwait, the Netherlands, New Zealand, Norway, Pakistan, Poland, Portugal, Qatar, Russia, Saudi Arabia, South Africa, South Korea, Spain, Taiwan, the United Arab Emirates, the UK, and the US. Concerns have also been raised over the potential of LAWS, although there are no fully autonomous systems to date. The overall global market for automated technology is predicted to reach US $20bn by 2025. The US and Israel are the two biggest developers of automated technology worldwide. There are also major emerging markets for unmanned systems in China, Russia, South Korea and India. The US Department of Defence claims that China is likely to produce 42,000 unmanned air and maritime systems during 2014–2023. Robotic armed stations have been deployed in South Korea for border surveillance and in 2011, Israel deployed its Iron Dome air defence system. Japan is the technological leader in UGVs, but has not been working on military systems because investments in that area are not lucrative. With the expansion of the global market for automated systems, these systems are becoming more affordable. A wide range of States and non-State groups are, therefore, likely to make use of automated technology in the future. In the future, LAWS will have even greater utility as they become faster, stealthier and more autonomous. They will become more accessible to foreign countries and terrorist groups around the world in the due course of time.

Endnotes

1   It has been proposed that a military–technical revolution is under way and a new warfare regime based on unmanned and autonomous systems has the potential to change our basic core concepts of military strategy. For more details, see; Work, Robert O., and Shawn Brimley, 20YY: Preparing for War in the Robotic Age, Centre for a New American Security, January 2014.

2   Ackerman E., US Army Considers Replacing Thousands of Soldiers with Robots, IEEE Spectrum, 22 January 2014.

3   Sapaty Peter Simon, Military Robotics: Latest Trends and Spatial Grasp Solutions, *International Journal of Advanced Research in Artificial Intelligence*, Vol. 4, No.4, 2015, pp. 9–18.

4   Peter Singer has characterized the levels of autonomy as "direct human operation", "human-assisted", "human delegation", "human-supervised", "mixed-initiative", "fully autonomous", and "adaptive". Singer, Peter W. *Wired For War: The Robotics Revolution and Conflict in the Twenty-First Century.* New York: Penguin, 2009, p. 74.

5   Due to the secrecy shrouding military technology, it is difficult to ascertain precisely the current cutting-edge capability of military robotics. Liu Hin-Yan, Categorization and legality of autonomous and remote weapons systems, *International Review of the Red Cross*, Vol. 94, No. 886, Summer 2012, pp. 627–652. For the US Department of Defence roadmap on unmanned systems, see: Unmanned Systems Integrated Roadmap FY2011-2036, available at: http://www.defenseinnovationmarketplace.mil/resources/ UnmannedSystemsIntegratedRoadmapFY2011.pdf, accessed 12 July 2015.

6   The regular (IED/EOD) TALON carries sensors and a robotic manipulator, which is used by the US military for explosive ordnance disposal and disarming improved explosive devices. Special Operations TALON (SOTAL) does not have the robotic arm manipulator but carries day/night colour cameras and

listening devices for reconnaissance missions. It is lighter due to the absence of the arm. The Special Weapons Observation Remote Direct-Action System (SWORDS) TALON is a UGV and used for small arms combat and guard roles. The HAZMAT TALON uses chemical, gas, temperature and radiation sensors that are displayed in real time to the user on a hand-held display unit. HAZMAT TALON robots use JAUS (Joint Architecture Unmanned Systems) software to make it possible to 'plug and play' up to seven detection devices mounted on a quick-release universal mounting tray for easy removal.

7 Providing supplies to troops in general is a dangerous operation due to the various ambush opportunities as supply troops typically move slower than combat troops. Logistic robot equivalents are also used in industry to manage heavy payloads to relieve human workers. Logistic robots can supply more cargo to the various hot spots, besides relieving human soldiers from carrying load, and thus making them more effective in theory (less fatigue, higher level of concentration). Struijk Bob, New Design Philosophy in Military Robotics, available at: http://www.repulestudomany.hu/folyoirat/2012_1/Bob_Struijk. pdf, accessed 10 July 2015.

8 The LS3 project has been funded by DARPA and the US Marine Corps. Boston Dynamics has assembled a team to develop the LS3, including engineers and scientists from Boston Dynamics, Carnegie Mellon, the Jet Propulsion Laboratory, Bell Helicopter, AAI Corporation and Woodward HRT. Marc Raibert, Kevin Blankespoor, Gabriel Nelson, Rob Playter, BigDog, the Rough-Terrain Quaduped Robot, available at: http://www.bostondynamics.com/ robot_bigdog.html, accessed 12 June 2015; US marines reject BigDog robotic packhorse because it's too noisy, available at: https://www.theguardian.com/ technology/2015/dec/30/us-marines-reject-bigdog-robot-boston-dynamics-ls3-too-noisy, accessed 27 May 2016.

9 Defence Review, Samsung SGR-A1 Armed/Weaponized Robot Sentry (or 'Sentry Robot') Remote Weapons Station (RWS) Finally Ready for Prime Time? 17 September 2014. Available at: http://www.defensereview.com/ samsung-sgr-a1-armedweaponized-robot-sentry-or-sentry-robot-remote-weapons-station-rws-finally-ready-for-prime-time/, accessed 22 July 2015.

10 Available at: http://sputniknews.com/world/20130610/181595843/India-Developing-Robot-Soldiers-Says-Research-Agency.html#ixzz3s8Sx1lXt, accessed 15 November 2015.

11 Libel Tamir and Boulter Emily, Unmanned Aerial Vehicles in the Israel Defence Forces: A Precursor to a Military Robotic Revolution, The RUSI Journal, Vol. 160, No. 2, April/May 2015, pp. 68–71.

12 The X-47B is likely to be integrated into the Unmanned Carrier-Launched Airborne Surveillance and Strike (UCLASS) system, which the US Navy

plans to field by 2020. UCLASS will greatly extend the reach of carrier strike groups, with a planned capability to fly two unrefuelled carrier orbits at 600 nautical miles out, or a single unrefuelled orbit at 1,200 nautical miles. The aircraft will be able to identify targets for manned aircraft or deliver limited strike capabilities in hostile airspace without putting pilots at risk at ranges exceeding that of manned aircraft. UCLASS is projected to have a small radar footprint and a 3,000-pound payload (including 1,000 pounds of air-to-surface weapons), which should make for improved carrier-based strike capability against anti-ship threats. The United States Naval Aviation Vision 2014–2025; Also see: Subramanioan Arjun, National Defence and Aerospace Power, Centre for Air Power Studies, Issue Brief, 89/13, 25 June 2013.

13 The Iron Dome comprises three key components: (i) the design and tracking radar, (ii) the battle management and weapon control system, and (iii) the missile firing unit. One of the most advanced features of the Iron Dome is its capability to determine where an incoming rocket will land and to only intercept such projectiles that pose meaningful threats to populated civilian areas. The Iron Dome is intended to protect against mid-range missiles, and the interceptor system, designed to provide defence against long-range ballistic missiles.

14 N. Marsh, 'Defining the Scope of Autonomy', PRIO Policy Brief 02, 2014.

15 Crane David, Israel Developing 'Bionic Hornet' to Target and Kill Enemy Combatants: Nanotechnology Goes Lethal. Available at: http://www. defensereview.com/israel-developing-bionic-hornet-to-target-and-kill-enemy-combatants/, accessed 12 July 2015.

16 The DRDO is the research and development arm of the Ministry of Defence. Its mandate is to provide assessment and advice on the scientific aspects of weapons, platforms and surveillance sensors; to carry out research; and to develop cutting edge technologies leading to the production of state-of-the-art sensors, weapon systems, platforms and allied equipment for our defence services. Ministry of Defence, Government of India, Annual Report 2014–2015, p, 72.

17 The lineage of sea-based robots dates to the nineteenth century when Robert Whitehead, a British engineer working for Austria in the 1860s, invented a gyro-stabilized torpedo propelled by a compressed air motor, making it the first unmanned underwater vehicle. During World War II, the US Navy developed the Mk 24 self-propelled mine, a sonar-equipped torpedo that rested on the sea bottom until it detected an enemy submarine. The sonar activated the weapon and guided it to its target, making it the first intelligent unmanned underwater vehicle. B. Berkowitz, Sea Power in the Robotic Age, *Issues in Science and Technology*, Vol. XXX, No. 2, 2014.

18   Some of these include: (i) improved navigation systems using sonar for higher-fidelity sea bottom mapping and detection of obstacles; (ii) miniaturized, solid-state inertial navigation units that allow vehicles to track their position with greater precision via dead reckoning; (iii) fibre optic communications, both to link ROVs to their parent ship and small ROVs to larger host vehicles; (iv) GPS, providing location data via either direct satellite links (in the case of USSs, and UUSs that periodically surface), or through a fibre optic link to a parent ship or buoy (in the case of ROVs); (v) acoustic modems that can transmit digital data through water, albeit at a lower rate and shorter ranges than their electronic counterparts, and acoustic positioning systems; (vi) side-scanning and synthetic aperture sonar, electro-optical cameras, magnetometers, and other sensors; and (vii) power innovations, such as lithium batteries and fuels cells.

19   The Weaponization of Increasing Autonomous Technologies in the Maritime Environment: Testing the Waters, The United Nations Institute for Disarmament Research (UNIDIR), UNIDIR Resources, No. 4, 2015.

20   White Andrew, Maritime Remotely Operated Vehicles (ROV) and Autonomous Underwater Vehicles (AUV), *Military Technology*, Vol. XXIX, Issue 6, 2015, pp. 86–87.

21   The US Navy has identified five major benefits of using modern unmanned vehicles in maritime surface and sub-surface applications. These are: (i) unmanned vehicles are far less expensive to operate and maintain than manned vehicles; (ii) automated sensors are able to maintain near-constant awareness and coverage of an environment; (iii) the near-constant surveillance permits persistency in data collection, allowing for a better understanding of long-term behaviour patterns and trends; (iv) unmanned platforms also promise to improve productivity, as they allow manned platforms to pursue tasks elsewhere; and (v) unmanned platforms keep human sailors and expensive manned platforms away from danger. The US Department of Defence, Department of the Navy, The Navy Unmanned Surface Vehicle (USV) Master Plan, 23 July 2007, p. 4.

22   The Weaponization of Increasing Autonomous Technologies in the Maritime Environment: Testing the Waters, The United Nations Institute for Disarmament Research (UNIDIR), No. 4, UNIDIR Resources, 2015.

23   The US Navy's current UUV Master Plan (2004) defines nine sets of UUV missions in the following prioritized order: (i) ISR, (ii) mine countermeasures (MCM), (iii) anti-submarine warfare (ASW), (iv) inspection/identification, (v) oceanography, (vi) Communications/Navigation Network Node (CN3), (vii) payload delivery, (viii) information operations, and (ix) time critical strike (TCS). The US Department of the Navy, *The Navy Unmanned Undersea Vehicle (UUV) Master Plan*, November 2004, p. 16. Also see: Button Robert

W., et. al., 2009, A Survey of Missions for Unmanned Undersea Vehicles, RAND Corporation, p. 3.

24 The US Office of Naval Research successfully tested a "swarm escort" of 13 unmanned surface vessels in 2014. See: Freedberg S., Naval Drones 'Swarm', But Who Pulls the Trigger, in *Breaking Defence*, 5 October 2014.

25 The US doctrine related to UMSs comes from three principal sources: The Unmanned Systems Integrated Roadmap FY2011-2036; The Navy Unmanned Undersea Vehicle Master Plan (2004); and The Navy Unmanned Surface Vehicle Master Plan (2007). The Navy Unmanned Undersea Vehicle Master Plan and The Navy Unmanned Surface Vehicle Master Plan are to some extent outdated. Norris Andrew, Legal Issues Relating to Unmanned Maritime Systems, Monograph, 2013, The US Naval War College, p. 5. The US Navy's master plan recommends the development of four classes of unmanned undersea vehicles (UUVs). The UUVs, from the smallest to the largest are: (i) the man-portable class, which includes vehicles that displace about 25–100 pounds and have an endurance of 10–20 hours, (ii) the light-weight vehicle (LWV) class, which is 12.75 inches in diameter and displaces about 500 pounds, (iii) the heavy-weight vehicle (HWV) class, which is 21 inches in diameter and displaces about 3000 pounds and includes submarine-compatible vehicles, and (iv) the large vehicle class, which displaces approximately 10 long-tons and is compatible with both surface ship and submarine use.

26 Parliamentary Questions, 18 November 2014, Tom Watson to Secretary of State for Defence, "The Royal Navy is, however, developing a maritime UAS strategy paper describing the requirement to 2050."

27 The Financial Express, http://www.financialexpress.com/article/economy/keeping-an-eye-underwater/60807/. Also see: http://www.indiandefensenews.in/2015/02/indian-navy-impressed-with-drdo-s.html, accessed 20 November 2015.

28 Committee for the Review of ONR's Uninhabited Combat Air Vehicles Program, Review of ONR's Combat Air Vehicles Program, Naval Studies Board.

29 The Weaponization of Increasing Autonomous Technologies in the Maritime Environment: Testing the Waters, The United Nations Institute for Disarmament Research (UNIDIR), No. 4, UNIDIR Resources, 2015, p. 7.

30 Berkowitz, B., Sea Power in the Robotic Age, *Issues in Science and Technology*, Vol. XXX, No. 2, 2014, available at: http://issues.org/30-2/bruce-2/.

IV   LAWS and International Law

International humanitarian law (IHL) is a branch of international law which limits the use of violence in armed conflicts. The basic principle of IHL is that in any armed conflict, the right of the parties to conflict to choose methods and means of warfare is not unlimited. IHL has developed over more than a century with a two-fold aim: to save civilians from the consequences of armed conflict, and to protect soldiers from cruelty and unnecessary suffering. The rapid advancement in autonomous technologies, in particular lethal autonomous weapons systems (LAWS), presents certain challenges to the basic tenets of IHL. Though there are international agreements to specifically ban or regulate a number of inherently problematic weapons, such as expanding bullets, poisonous gases, antipersonnel landmines, biological and chemical weapons, blinding lasers, incendiaries, and cluster munitions, there is no regime for LAWS.

The Latin phrases *jus ad bellum* and *jus in bello*[1] describe the law governing resort to force and the law governing the conduct of hostilities. These are recognized branches of international law and are generally independent. The morality and legality of a state deciding to go to war (*jus ad bellum*) is something that is decided by a state's political leadership. LAWS are likely to present complex *jus ad bellum* issues related to lowering the moral, political and financial cost of warfare for those States that have the capability to develop and deploy them in an armed conflict. The use of LAWS in armed conflict or in enforcement operations could be incompatible with international human rights law, and may lead to unlawful killings, injuries and other violations of human rights.

This chapter will address the issue of *jus ad bellum* proportionality, and three key concerns related to LAWS: their implications for the principles of distinction, proportionality and precautions in attack. It will also discuss the challenge the posed by LAWS to accountability and enforcement. The likely impact of the recently adopted Arms Trade Treaty (ATT) in regulating LAWS will also be covered in this chapter. The human rights implications of LAWS will be dealt with in the last part of the chapter.

Jus Ad Bellum Proportionality

Jus ad bellum comprises six principles: just cause, right intention, proper authority, last resort, the probability of success, and the response of declaring war being proportionate. [2] The principle of 'just cause' relates to the normative reasons for waging war, such as self-defence or defence of others, while 'right intention' prescribes the proper reasons for acting. The principle of 'proper authority' dictates that only legitimately recognized authorities may declare a war. Last resort requires states to attempt all reasonable alternatives available, such as diplomacy or arbitration before resorting to hostilities. Probability of success requires states to assess whether an actor is able to achieve its just cause through fighting a war. The principle of proportionality dictates that we must consider the overall consequences of a proposed war.

LAWS pose a distinct challenge to the *jus ad bellum* proportionality principle. Proportionality is closely linked to other *ad bellum* principles and it requires that we consider the overall consequences of a proposed war. If we cannot satisfy the principle of proportionality, the other principles can never meet the obligations of a just war.[3] A State that has LAWS in its arsenal will have advantage of using them in defence, particularly in cases where the other side lacks the same level of technology.[4] LAWS saves soldiers' lives and this would weigh heavily in favour of deploying such weapons, for the harms caused would be blamed on the unjust aggressor and not on the defending State. In case LAWS cause collateral harm, the State could justify it by declaring that it was unintended and that LAWS were used in pursuance of legitimate military objectives. The ability to use LAWS against an unjust threat must be seen as a benefit in one's proportionality calculation.[5]

The presence of LAWS might influence the choice of a nation to go to war in two ways: (i) it could directly threaten the sovereignty of a nation, and (ii) it could make it easier for leaders who wish to start a conflict to actually start one. In other words, the availability of LAWS would lower the barrier to initiate an armed conflict. The presence of LAWS is likely to favour enhanced potential for armed deployments. The current use of armed drones by the US amplifies this. It is thought that future LAWS would be capable of learning and adapting their functioning in response to changing circumstances in the environment in which they are deployed, as well as be capable of making firing decisions on their own. Such systems could be directly responsible for starting an armed conflict accidently. The use of LAWS in armed conflict is also likely to increase the arms race. [6]

Distinction

Distinction is one of the most important principles of IHL. It requires combatants to direct their attacks solely at other combatants and military targets and to protect civilians and civilian property.[7] Under the principle of distinction, indiscriminate attacks are prohibited. Indiscriminate attacks are those that are not directed at a military object,[8] or employ a method or means of combat the effects of which cannot be directed or restricted as required. The principle of distinction also necessitates that defenders must distinguish themselves from civilians and refrain from placing military personnel or material near civilian objects.[9] A major IHL issue is that LAWS cannot discriminate between combatants and non-combatants or other persons likely to be present at the place of conflict. The list includes civilian workers, aircrew members, war correspondents, military doctors, religious personnel, civilian drivers, porters as well as combatants who are unwilling to fight or are wounded or sick.

According to Sharkey (2012) LAWS lack three of the main components required to ensure compliance with the principle of distinction: (i) adequate sensory processing systems for distinguishing between combatants and civilians; (ii) programming language to define a non-combatant or person *hors de combat*; and (iii) battlefield awareness or common sense reasoning to assist in discrimination decisions. Even if LAWS have adequate sensing mechanisms to detect the difference between civilians and combatants, they would lack 'common sense' which is used by an experienced soldier on the battlefield for taking various decisions.[10] At present there is no evidence to suggest that a computer has independent capability to operate on the principle of distinction similar to that of a human soldier.

Due to the vagueness of the legal definitions contained in the Geneva Conventions of 1949 and AP I, it is not possible to incorporate the essence of the principle of discrimination into the programming language of a computer. A human combatant has to take positive steps to understand a situation; develop his/her own mental model of offence or defence; and then recommend or demand engagement, a process that is extremely difficult to incorporate in LAWS. Even if we equip LAWS with mechanisms to distinguish between civilians and military combatants, these devices lack the capacity to reach the human level of common sense that is indispensable for the correct application of the principle of discrimination. The application of the principle of proportionality is more difficult than distinction since it involves comparing an action's potentially excessive collateral damage to its anticipated military benefits. It requires a case-by-case strategic and military evaluation, which a machine simply cannot comprehend.[11]

Sparrow has highlighted an important shortcoming of LAWS, i.e., the question of the capacity of LAWS to recognize surrender, and the implication of this for the ethical deployment of such weapon systems.[12] A fundamental requirement of the principle of distinction is that combatants should not attack enemy units that have clearly indicated their desire to surrender.[13] By ceasing to participate in hostilities and signalling surrender, military units can acquire the moral status of non-combatants, such that deliberate attacks on them are no longer permissible.

It can be argued theoretically that if a combatant can recognize a signal to surrender then it should not be impossible for LAWS to do so. However, Sparrow has advanced two reasons why recognizing surrender is likely to be difficult for robots. The first relates to the fact that 'perception' would be a formidable task for computers. In spite of recent developments in robotic technology, real-time recognition of objects in motion across a range of environments remains beyond the capacity of even the most sophisticated computer vision systems.[14] The second relates to the contextual nature of the means used to signal surrender in different circumstances and the problem of distinguishing between surrender and perfidy. Human beings have a tremendously sophisticated and powerful capacity to interpret the actions of other human beings and to identify their intentions. It will be extremely challenging to design a machine that comes close to replicating these capabilities.[15]

It may be argued that though an autonomous system might be unlawful because of its ability to distinguish civilians from combatants in the operational conditions of infantry urban warfare, for example, it may be lawful in battlefield environments with very few if any civilians present.[16] However, the casualty figures of the post-World War II armed conflicts indicate that an increasing number of civilians are becoming victims of modern warfare. Therefore, LAWS may not be suitable weapons of future armed conflict.

Proportionality

The principle of proportionality is related to the principle of distinction. It prescribes that belligerent parties in an armed conflict are not to inflict collateral damage that is excessive in relation to the military advantage they seek with any hostile action. This principle is considered part of customary international law, which binds all states.[17] It provides that an attack which may be expected to cause incidental loss of civilian life, injury to civilians, damage to civilian objects, or a combination thereof, which would be excessive in relation to the concrete and direct military advantage anticipated is prohibited.[18] In order to make the principle of proportionality more effective, Article 57 of AP I obliges belligerents to take all feasible precautions and constant care with a view to implementing proportionality and distinction.[19] If damage to civilian objects or civilian death or injury is anticipated prior to targeting a military object, then an assessment must be undertaken in which the anticipated military advantage to be gained is weighed against the anticipated "collateral" damage to protected civilians or civilian objects. Viewed from this perspective, LAWS, which are incapable of making a distinction between combatants and civilians, would not be able to follow the principle of proportionality.

Arkin (2009), suggests that it is possible to develop proportionality optimization algorithm for LAWS. The algorithm would select the weapon system that ensures that it would not violate any proportionality prohibitions or IHL. It would calculate the potential unintended noncombatant carnage and civilian property damage (collateral damage) that would result from available combinations of weapon systems and release positions, choosing the most effective weapon that would cause the lowest acceptable collateral damage.[20] On the other hand, Human Rights Watch has serious doubts about whether LAWS could exercise comparable judgment to assess proportionality in complex and evolving situations. It would be difficult to programme LAWS to carry out the proportionality test or to visualize every situation of armed conflict because there are an infinite number of possible situations.[21]

Professor Sharkey is of the opinion that though it might be possible for LAWS to be programmed to observe the principle of proportionality in a limited way, or to minimize collateral damage by selecting appropriate weapons and properly directing them, it would not be possible to guarantee respect for the principle of proportionality in the near future. Only a human being can make qualitative and subjective decisions on when damage to civilians would exceed the anticipated military advantage provided by an attack.[22] Kastan is of the firm belief that whatever technological advances may come about, the relevant analysis and assessment of the principle of proportionality would have to be left to human beings. [23]

The US Air Force maintains that proportionality in attack is an inherently subjective determination that could be resolved on a case-by-case basis.[24] In reality, it would be nearly impossible to pre-programme LAWS to handle the infinite number of scenarios it might face in a fog of war. It would therefore be extremely difficult to programme a machine to replicate the decision-making capabilities of a military commander or a combatant. Therefore, non-compliance with the principle of proportionality, in addition to failure to distinguish between civilians and combatants, could lead to an unlawful loss of innocent lives.[25] A breach of the rule of distinction or proportionality is regarded as a serious violation of law of armed conflict. It is listed as a grave breach of the 1977 AP I and as a war crime under the 1998 Rome Statute of the International Criminal Court.[26]

It would be very challenging to program a machine to: measure anticipated civilian harm and measure military advantage; subtract and measure the balance against some determined standard of "excessive"; if excessive, not to attack an otherwise lawful target. From a programming point of view, this would require attaching values to various targets, objects, and categories of human beings, and making probabilistic assessments based on many complex contextual factors. It may also include inductive machine learning from human examples of judgments about proportionality, seeking to extract practical heuristics from them.[27] It may be possible for engineers and programmers to do this in future, but today there is no possibility of developing such autonomous systems. The States which participated in the first UN Expert Meeting of CCW in May 2014 recognized respect for IHL as an essential condition for the implementation of LAWS. With diverse predictions, some States believed that LAWS would be unable to meet the IHL criterion, while a few expressed difficulty in reaching a conclusion without knowing the weapons' future capabilities. Organizations and individuals in favour of a complete ban on LAWS are of the opinion that no matter what technological progress takes place, machine systems will never reach the point of satisfying the legal or moral requirements of IHL.

Precautions in Attack

International humanitarian Law provides that in the conduct of military operations, whether on land, at sea or in the air, the parties to a conflict must take all reasonable precautions to avoid loss of civilian lives and damage to civilian objects. The parties to a conflict are under an obligation to spare the civilian population and civilian objects from the effects of armed conflict. This obligation covers actions in both offence and defence and applies to all personnel; even an act of a single solider in attack could be covered. For instance, a pilot who is on a bombing mission is required to meet this obligation. Likewise, those who plan or decide upon an attack must do everything feasible to verify that the objectives to be attacked are military objectives and are not under any form of protection.[28] What is 'feasible' would depend upon the resources and technology available with a commander who is planning an attack.[29] This customary obligation of taking 'feasible precautions' can be justly expected from a military commander in conventional warfare. However, in an armed conflict in which LAWS are deployed, it would be nearly impossible for such systems to take all 'feasible precautions'.[30] While it may be possible for a commander to consider updated information available from the battlefield, a machine cannot be programmed with every futuristic scenario of an armed conflict.

The ICRC Commentary provides that a commander planning an attack must in case of doubt (even if there is only a slight doubt), call for additional information and if need be give orders for further reconnaissance.[31] In case of long-distance attacks, information may be gathered with the help of aerial reconnaissance and intelligence units. Before planning an attack, the information so obtained must be checked for its accuracy by the commander. The problem with the use of LAWS is that such a system would not be infused with the latest inputs from the surrounding environment, while deciding to use lethal force. While a military commander may cancel or suspend an attack, such possibility may be ruled out with the deployment of LAWS. The lack of multiple intelligence sources could also inhibit the ability of LAWS to identify targets accurately. Developing LAWS with an 'intelligent system' that is similar to or better than 'human' is perhaps not feasible in the near future.

Responsibility and Accountability

As technology advances towards the creation of autonomous military systems, lawyers and military analysts are debating the issue of responsibility and liability in the event of such weapon systems being used. Since no international conventions directly regulate autonomous military systems, the question is that who should be held responsible when LAWS carry out strikes in violation of IHL or IHRL? The mere fact that a human is not in control of a particular weapon system does not mean that no human is responsible for the actions of the system.

At present, unmanned ground, air, and underwater systems are operated remotely by humans, meaning that human operators are still 'in-the-loop'. A few countries are developing autonomous weapons systems, some of which are in use, though these cannot be categorized as LAWS in the true sense. For instance, South Korea's SGR-A1 sentry robots have a fully autonomous mode, which would allow the robots to make the decision to shoot. Israeli Sentry Tech systems deployed along the Gaza border also have an autonomous firing mode; the Guardium, which patrols the border, can respond to stimuli from the environment; and the US Navy's X-47B has the capability to take off from, land on and refuel at an aircraft carrier without human supervision.

During an armed conflict, soldiers are capable of making decisions and reflecting on their own autonomy. Thus they can be held responsible for any action they choose to undertake. They can be prosecuted for violating the principles and laws of IHL. Autonomous weapons, on the other hand, lack the capacity for moral autonomy, and thus cannot be held responsible. Even if one supposes that some LAWS are capable of discriminating between combatants and noncombatants, it holds no moral weight. Since LAWS are not moral agents, there can be no moral equality between them and soldiers on a battlefield. The only way equality can be maintained is if both sides involved in a conflict decide to employ LAWS.

We hold political and military leaders morally, and sometimes legally, responsible for armed conflict. We have tried political and military leaders as well as individual soldiers in domestic and international courts. We praise them for conducting a successful armed conflict and blame them for committing serious violations of IHL or of human rights. However, holding political or military leaders accountable for the actions of LAWS would pose a number of problems.

Fixing of Responsibility

Several people with varied expertise would play a role in the development of LAWS.[32] The military commander may not play a significant role in the deployment and use of LAWS in an armed conflict. It would, therefore, be extremely difficult to establish accountability in the case of an unlawful act caused by LAWS.

Reparations to the Victims of LAWS

International law prescribes that States must bear responsibility for illegal acts attributable to them and that such responsibility entails a duty to provide reparation.[33] The Basic Principles and Guidelines on the Right to a Remedy and Reparation for Victims of Gross Violations of International Human Rights Law and Serious Violations of IHL, adopted by the UN General Assembly in 2005, recognize the importance of the international legal principles of accountability, justice and the rule of law and lay down the elements of an accountability regime.[34]

The right under the Basic Principles includes access to justice, reparation for harm suffered and access to factual information concerning the violations. The Basic Principles distinguish between five forms of reparation: restitution, compensation, rehabilitation, satisfaction, and guarantee of non-repetition. They focus on victims of violations of IHL and define a victim in the following terms: A person is 'a victim' where, as a result of acts or omissions that constitute a violation of international human rights or humanitarian law norms, that person, individually or collectively, suffered harm, including physical or mental injury, emotional suffering, economic loss, or impairment of that person's fundamental legal rights. The reparation should be provided by the state responsible for the violation.

The purpose of assigning personal responsibility is to deter future violations and to provide reparation to victims. Such accountability under IHL imposes a duty to prosecute war crimes. The Rome Statute of the International Criminal Court authorizes the Court to determine any damage, loss or injury to victims and order reparations to them. The 2005 ICRC Rules of customary IHL provide, "A State responsible for violations of international humanitarian law is required to make full reparation for the loss or injury caused."[35] In addition, recent IHL treaties[36] also impose obligations on States to provide reparations to the victims. State responsibility brings a legal duty to make reparations for violations. The general rule on reparation for violations of international law was stated in the Permanent Court of International Justice's 1928 decision in the *Chorzow Factory case*: "It is a principle of international law, and even a general conception of the law, that any breach of an engagement involves an obligation to make reparation. ...Reparation is the indispensable complement of a failure to apply a convention, and there is no necessity for this to be stated in the convention itself." The Inter-American court of Human Rights has adjudicated a number of cases involving gross violations of human rights, including killings. The Court has ruled that Article 25 of the American Convention on the rights to judicial protection and effective domestic recourse is one of the fundamental pillars not only of the American Convention, but of the very rule of law in a democratic society in term of the Convention.[37] Therefore if the uses of LAWS causes illegal death, physical or psychological sufferings or injury, or economic losses, the State employing such weapons must provide reparations to victims/ their families.

A. State Accountability and Liability

In international law, States are responsible for their military and other agents and thus could be liable for all violations of IHL.[38] A State acquiring, developing, or using LAWS would be under an obligation to ensure that its use does not violate IHL or human rights. It could be held liable for war crimes committed by LAWS deployed by it.[39] The State, while implementing and upholding IHL, is also under an obligation to investigate and to punish violations of IHL committed by its agents. For gross violations of Geneva Conventions of 1949 or provisions of 1977 AP I, legal process could take place in national courts or else the International Criminal Court would have jurisdiction over such crimes. The State is also under an obligation to pay reparations for the gross violations of human rights and serious violations of IHL. Article 91 of AP I also provide that a party who commits violations shall be liable to pay compensation if the case demands.[40] However, the main obstacle in holding a state responsible for war crimes committed by LAWS would be whether the act was intentional or negligent. In today's political scenario, the liability and reparation, if any, would be dependent on the State to which the victims belong.[41] An individual victim may be too poor to legally confront an offending state.[42]

B. Individual Responsibility

As software becomes more complicated, future LAWS may become less predictable. No one programmer will understand or know the entire piece of software, so interactions within it will be unpredictable as well. Combined with an open environment, this could lead to situations where LAWS may apply force indiscriminately because of an unanticipated software error.[43] With a long list of individuals involved in programming and manufacturing, the use of LAWS in a battlefield may raise questions on individual responsibility for a wrong committed by an errant machine. The list of individuals who could be held responsible could include the combatant, the military commander, the programmer/s, the manufacturer/s, the corporation, the State or possibly the LAWS itself; or no one?

I. Whether LAWS could be blamed

A few authors have suggested that where a machine operates with a degree of independence, it could be granted legal personality and could be held accountable for violations of the law.[44] However, while applying the rules of IHL, some preliminary issues must be clarified. Only human beings are subject to legal rules and only human beings are obliged to follow them. In the case of LAWS, IHL could be applicable to those who devise them, produce them, programme them, and decide on their use. No matter how far we go into the future and regardless of the ways in which LAWS work, there will always be a human being involved, at least in the machine's conception. A human being will decide to create this machine and will then create the machine. Even if one day LAWS construct LAWS, it will still be a human being who will construct the original machine. This human being is bound by the law. The machine is not bound by the law.[45]

II. Command Responsibility

The autonomous nature of weapons would make them legally analogous to human soldiers and could trigger the doctrine of indirect command or superior responsibility. The view that a superior could be held criminally liable in relation to crimes committed by subordinates was applied in several cases after the Second World War. Presently contained in Article 87 of the 1977 AP I and Article 28 of the Rome Statute, command responsibility holds superiors accountable only if they knew or should have known of a subordinate's criminal act and failed to prevent or punish it. To be held criminally responsible as a superior for a failure to prevent or punish crimes of subordinates, the accused must be shown to have been in a superior-subordinate relationship with those who committed the crimes which form the basis of the charges against him.[46] These criteria set a high bar for accountability for the actions of a fully autonomous weapon.

Command responsibility deals with the prevention of a crime, and since machines cannot have the mental state to commit an underlying crime; command responsibility can never be available in situations involving LAWS. LAWS are embedded with complicated technology and are designed to operate independently, so a commander can never be held accountable for alleged violations of IHL by such machines. How can the commanding officer be held responsible for the actions of LAWS, if such weapons choose their own targets and kill innocent people? [47] According to Special Rapporteur Heyns, it remains unclear whether military commanders will be in a position to understand the complex programming of LAWS sufficiently well to warrant criminal liability. In addition it would always be doubtful whether LAWS could be considered a subordinate to a military commander.

C. Developer/ Manufacturer's Liability

Due to the complexity of the process of development, it will not be easy even for those involved in the manufacture of LAWS to estimate all the possible consequences of their deployment.[48] According to a Human Rights Watch report, it would not be possible to hold the manufacturers liable for any harm caused, if: (i) the specification for LAWS was approved by the government, (ii) the weapon conformed to those specifications, and (iii) the manufacturer did not deliberately fail to inform the government of any expected or known danger from the weapon system. Besides, it may be difficult to prove that the affected person was harmed due to the failure of LAWS or because the system did not function according to the approved design. According to Sparrow, holding the programmers or manufacturers responsible for the actions of their creation, once it is autonomous, "would be analogous to holding parents responsible for the actions of their children once they have left their care".[49]

The major disadvantage of civil law proceedings is that they impose an obligation on victims to make a complaint. Given the nature of conflict situations today, potential victims of LAWS would rarely be in a position to initiate proceedings against the manufacturers. The developers and manufacturers would most likely be located in industrialized countries and filing a civil suit against them would be next to impossible.[50] Therefore they would most likely escape liability for their role in providing LAWS to the armed forces.[51]

The US Department of Defence has emphasized that "persons who authorize the use of, direct the use of, or operate autonomous and semi-autonomous weapon systems must do so with appropriate care and in accordance with the law of war, applicable treaties, weapon system safety rules, and applicable rules of engagement (ROE)."[52] However, whether anyone is held accountable or punished suitably by the US military justice system for violations of IHL or human rights by autonomous weapon systems is yet to be seen. Scholars and NGOs firmly believe that it would be impossible under existing international law to hold someone accountable and liable for violations of IHL by LAWS.[53]

The development and the deployment of LAWS could mean a step backward for international criminal law. The use of LAWS is likely to create a legal vacuum in the sphere of personal legal responsibility for war crimes committed by such weapons.[54] A few scholars have suggested that the inability to hold someone accountable for war crimes committed by LAWS could be one of the reasons for banning their development.[55] Many of the risks, dangers, and challenges of future LAWS are already present in existing semi-autonomous systems like lethal drones that are widely deployed.[56]

Arms Trade Treaty

The global trade in conventional arms and ammunition fuels conflict and human rights abuses. It causes massive displacement, undermines sustainable development and contributes to poverty. The lack of transparency and corrupt practices in the arms trade undermine the rule of law and good governance. Lack of accountability of States for the end-use of exported arms facilitates diversion of arms to the illicit market. The 2013 UN Arms Trade Treaty (ATT) aims to establish the highest possible common international standards for regulating the international trade in conventional arms and to eradicate the illicit trade in conventional arms for the purpose of contributing to international and regional peace, security and stability. The treaty regulates the trade among states in conventional arms, including battle tanks, combat aircraft, artillery, rockets, missiles, and small arms, as well as the trade in ammunition and parts and components of arms.

Background: In 1991, the five permanent members of the UN Security Council agreed on guidelines for "Conventional Arms Transfers", a set of criteria upon which they would base their arms export decisions. In the same year, the UN General Assembly adopted a resolution establishing the UN Register of Conventional Arms to promote transparency in the trade of conventional weapons.[57] In 1996, the UN Disarmament Commission adopted guidelines for international arms transfers. However, the guidelines and resolution were of non-binding nature and participation in the Register was not very encouraging. In 1997, a group of Nobel Peace Laureates published an International Code of Conduct on Arms Transfers, which was developed into a Framework Convention on International Arms Transfers in 2001. In the following years, a network of international NGOs initiated the Control Arms campaign advocating for an arms trade treaty. The United Nations took nearly a decade to react on the issue of regulations of arms trade.[58]

The ATT is the first international legally binding agreement to establish standards for regulating the trade in conventional arms and preventing the illicit trade in weapons. It was adopted by the UN General Assembly on 02 April 2013 and entered into force on 24 December 2014.[59] The treaty establishes a new global norm and provides international prohibitions on the sale and transfer of conventional weapons to armed groups that abuse human rights and humanitarian law, engage in organized crime or commit acts of terrorism and piracy. The ATT covers a wide range of conventional arms within the following categories: battle tanks; armoured combat vehicles; large-calibre artillery systems; combat aircraft; attack helicopters; warships; missiles and missile launchers; and other small arms and light weapons.[60]

The ATT requires all states to adopt basic regulations and approval processes for the flow of weapons across international borders, establishes common international standards that must be met before arms exports are authorized, and requires annual reporting of imports and exports to a treaty secretariat. In particular, the treaty requires that States should establish and maintain a national control system, including a national control list and designate competent national authorities in order to have an effective and transparent national control system regulating the transfer of conventional arms. The treaty prohibits arms transfer authorizations to States if the transfer would violate "obligations under measures adopted by the United Nations Security Council acting under Chapter VII of the Charter, in particular arms embargoes" or under other "relevant international obligations" or if the state "has knowledge at the time of authorization that the arms or items would be used in the commission of genocide, crimes against humanity, grave breaches of the Geneva Conventions of 1949, attacks directed against civilian objects or civilians protected as such, or other war crimes".[61]

In Article 7, the treaties requires States to assess the potential whether the arms exported would contribute to or undermine peace and security; or could be used to commit or facilitate serious violation of IHL or human rights law; or act of terrorism or transnational organized crimes under international conventions or protocols to which the exporting State is a party. If an "overriding risk" of "negative consequences" remains after the State considers measures to mitigate the risk of these violations, the treaty requires that the State "shall not authorize the export".[62] The treaty does not provide guidance on determining what threshold the risks must meet to be considered "overriding". It requires States to establish and maintain a national control system to regulate the (i) export of ammunition/ munitions fired, launched or delivered by conventional arms, and (ii) the parts and components that could be assembled as listed in Article 2(1). Article 13 of the ATT obliges each state to provide the Secretariat, by the end of May each year, with a report for the preceding calendar year concerning authorized or actual exports and imports of conventional arms. The treaty allows States to exclude commercially sensitive or national security information.

The treaty is expected to work in favour of human rights and human security considerations when States consider arms sales in the future. It achieves a balance between the interests of importing and exporting States and places obligations on importing and exporting States, as well as transit States. As more States, sign, ratify, and implement the ATT, the current loopholes that enable illicit and irresponsible arms sales may be plugged.

ATT and LAWS

The ATT does not cover non-lethal autonomous systems which are used exclusively by the armed forces for military purposes.[63] Article 2, paragraph 3 specifies that the ATT does not apply to the international movement of conventional arms by a state or on its behalf (for example by a company), where the movement of arms is for that state's own use, and while the arms concerned remain under that states ownership. Therefore, if a State sends autonomous systems abroad to its own forces, no transfer occurs. Also, if the ownership of such systems is then passed on to a non-State actor, such transfer would not contravene the provisions of the treaty. The ATT does not have provisions for regulating transfer of technologies.

The use of unmanned aerial systems by the armed forces is on the rise. The militaries of the developed nations are using a wide range of autonomous technologies on land, air and sea. These systems can be armed by the importing states. For example, TALON, a lightweight, unmanned, tracked military autonomous system was initially developed to protect combatants and detect and destroy explosive threats. Its large, quick-release cargo bay accommodates a variety of sensor payloads, making TALON a one robot solution to a variety of mission requirements. It can be modified to incorporate a weapons delivery system. The Special Weapons Observation Reconnaissance Detection System (SWORDS) is an armed version of TALON, carrying a machine gun or a grenade launcher for defence purposes rather than offensive actions. Such autonomous systems are not subject to ATT. However the can be fitted with weapon attachments that can be sold separately and reassembled later. Thus these unmanned surveillance systems can be bought as civilian systems and be adapted for military purposes later.[64] With an increasing global market for unmanned technology, distinctions between civilian and military applications are getting blurred and future conflicts are likely see an increasing use of these weapon systems.

There is a serious lacuna in the treaty as Article 6 of the ATT does not prohibit arms transfers to non-state actors. Article 7 (1) (a) requires states to assess whether the arms or items to be transferred could contribute to or undermine peace and security. The reference to a possible contribution to peace and security has been a contentious issue during the ATT conferences. Read with Article 7 (3) of the ATT, it could lead states to disregard the risk of the usage of exported weapons or items for violations of international law due to their possible contribution to peace and security under Article 7 (1) (a) of the treaty.[65]

The ATT has the potential to strengthen transparency and responsibility in arms transfers and to improve national and international weapons control capacities, and ultimately to decrease illicit arms trade, insecurity and corruption. However, for the treaty to complement and strengthen the existing instruments effectively, States and implementing bodies need to take steps to reduce clashes or overlaps, and explore synergies with related instruments and measures. The States could take into account the possible participation of the end users in the International Code of Conduct for Private Security Service Providers.[66] Future ATT deliberations should address the issue of interactions between the treaty and instruments to combat terrorism and piracy and establish means to enhance the controls on the arms-related actions of private military companies.[67] It is doubtful that the States will voluntarily apply the ATT's provisions to armaments other than those listed in Article 2(1) of the ATT.

Today, a number of developed states are involved in military conflicts driven by clashes with nondemocratic countries or interventions in unstable regions of the world. The irony is that the citizens of these countries want these adventures embarked upon by their governments to be low cost and with no casualties from their own military, but usually have no concern about the casualties inflicted on the enemy and local population. Unmanned systems for surveillance are already deployed by about 40 countries. Arming them with weapons is the next step.[68] If unchecked by preventive arms control, this process will spread to many more countries, as the existing laws of IHL are not capable of controlling the misuse of autonomous systems. Lastly, the non-signatories to the treaty, like Russia and China, are not under any obligations to follow the norms of the ATT.

Human Rights Implications

Autonomous weapon systems are capable of using lethal force against individuals and could have adverse consequences on a person's human rights. While highly effective in military operations, LAWS may harm the civilian population due to their speed, effectiveness and inability to distinguish between civilians and combatants.[69] They could also fall into the hands of non-state armed groups, criminal gangs and private companies and individuals. Their use in law enforcement would potentially affect a number of human rights, including the right to life, the right to dignity, the rights to liberty and security, and the prohibition of torture and other forms of cruel, inhuman, or degrading treatment.[70]

A. The Right to Life

Human rights law applies to the use of force at all times;[71] it is complementary to IHL during armed conflict, and where there is no armed conflict it applies to the exclusion of IHL. Therefore the use of LAWS during an armed conflict,[72] anti-terrorism operation or domestic law enforcement is clearly subject to international human rights law.

The right to life, which is a fundamental rule of international human rights law, can never lawfully be restricted, even in times of war or emergency. The Preamble to the Universal Declaration of Human Rights (UDHR) states that the recognition of the inherent dignity and of the equal and inalienable rights of all members of the human family is the foundation of freedom, justice and peace in the world. Article 3 of UDHR upholds the right of everyone "to life, liberty and security of person."[73]

Article 6 of the International Covenant on Civil and Political Rights (ICCPR) states: "Every human being has the inherent right to life. This right shall be protected by law. No one shall be arbitrarily deprived of his life." The regional human rights treaties from Africa, the Americas, and Europe have also incorporated the right to life.[74] The Human Rights Committee (HRC), in its General Comment 6, describes the right to life as the supreme right because it is a prerequisite for all other rights. It is non-derogable even in public emergencies that threaten the existence of a nation.

B. The Right to Life in Law Enforcement Situations

The right to life prohibits arbitrary killing. The use of potentially lethal force in law enforcement or internal armed conflict is governed by the principles of necessity and proportionality. These principles translate into the rule requiring the State's forces to effect an arrest where possible, as well as to plan their operations in such a way as to maximize the opportunity of being able to effect an arrest.

In this context, international human rights law stipulates that force must be used in a manner that safeguards the right to life, the right to dignity, the right to liberty, and prohibits torture and cruel, inhuman, or degrading treatment. In its General Comment 6, the Human Rights Committee highlights the duty of States to prevent arbitrary killings by their security forces. A number of criminal justice instruments govern the use of lethal force in law enforcement. They include the 1979 Code of Conduct for Law Enforcement Officials, and the 1990 Basic Principles on the Use of Force and Firearms by Law Enforcement Officials. Adopted by a UN congress on crime prevention and the General Assembly respectively, these standards provide important guidance for protecting human rights and the duties of the law enforcement officials. [75]

The 1979 Code provides that law enforcement officers include all officers of the law, whether appointed or elected, who exercise police powers, especially the powers of arrest or detention. In countries where police powers are exercised by military authorities, whether uniformed or not, or by State security forces, the definition of law enforcement officials shall be regarded as including officers of such services. In the performance of their duties, law enforcement officials are to respect and protect human dignity and uphold human rights; they may use force only when strictly necessary and to the extent required for the performance of their duty. [76]

The 1990 Basic Principles provides that the development and deployment of non-lethal incapacitating weapons should be carefully evaluated in order to minimize the risk of endangering uninvolved persons, and the use of such weapons should be carefully controlled. The law enforcement officials, as far as possible, shall apply non-violent means before resorting to the use of force and firearms. They may use force and firearms only if other means remain ineffective or without any promise of achieving the intended result.[77] Further, whenever the lawful use of force and firearms is unavoidable, law enforcement officials shall:

(a) Exercise restraint in such use and act in proportion to the seriousness of the offence and the legitimate objective to be achieved;

(b) Minimize damage and injury, and respect and preserve human life;

(c) Ensure that assistance and medical aid are rendered to any injured or affected persons at the earliest possible moment;

(d) Ensure that relatives or close friends of the injured or affected person are notified at the earliest possible moment.[78]

Law enforcement officials shall not use firearms against persons except in self-defence or defence of others against the imminent threat of death or serious injury, to prevent the perpetration of a particularly serious crime involving grave threat to life, to arrest a person presenting such a danger and resisting their authority, or to prevent his or her escape, and only when less extreme means are insufficient to achieve these objectives. In any event, intentional lethal use of firearms may only be made when strictly unavoidable in order to protect life. The Basic Principles also stipulates that Governments shall ensure that arbitrary or abusive use of force and firearms by law enforcement officials is punished as a criminal offence under their law. Exceptional circumstances such as internal political instability or any other public emergency may not be invoked to justify any departure from these basic principles.

The lawfulness of the use of lethal force during internal conflicts and law enforcement duties requires compliance with the principles of 'legitimate purpose', 'strict necessity' and 'proportionality of the force' used, and will need to be assessed on the facts of each individual case. Necessity is the first precondition for lawful force. The 1979 Code of Conduct states that law enforcement officials may employ force only when it is strictly necessary and exceptional. Fully autonomous weapons would lack human qualities that help law enforcement officials assess the seriousness of a threat and the need for a response.

Fully autonomous weapons would face obstacles in meeting these criteria that circumscribe lawful force. They would not be able to replicate the ability of human law enforcement officials to exercise judgment and compassion or to identify with other human beings, qualities that facilitate compliance with the law. Due to this inadequacy, the weapons could contravene the right to life, undermining the legitimacy of law enforcement operations and cause serious harm. In addition, the deployment of fully autonomous weapons in law enforcement situations could affect the actions of the individual, posing a potential threat. He or she might not know how to respond when confronted with a machine rather than a human law enforcement officer. The individual might respond differently to robotic weapons which may result in arbitrary killing.

There are serious doubts about whether fully autonomous weapons could determine how much force is proportionate in a particular case. A designer cannot pre-program a robot to deal with all situations because even a human being cannot predict the infinite possibilities. Law enforcement officials have suitable training and experience, coupled with subjective thinking to function in varied environments. A fully autonomous weapon would not be able to read a situation well enough to strategize about the best alternatives to the use of force. According to Christof Heyns, while robots are especially effective at dealing with quantitative issues, they have limited abilities to make the qualitative assessments that are often called for when dealing with human life. Allowing machines to determine whether to act in the defence of others would pose grave risks to the right to life. [79]

Further, if there is no immediate threat to life, a criminal must be afforded the rights under Article 14 of the ICCPR.[80] As a suspect, he has the right to a fair and public hearing by an impartial tribunal, to be presumed innocent, to have time and facilities to prepare a defence, to have the assistance of a counsel, to appear at a trial, to examine witnesses, and to appeal a judgment. The use of autonomous weapons to identify and kill a suspect would amount to having a preprogrammed machine play the role of judge, jury and executioner. Thus the use of such a weapon system would violate all the rights contained in Article 14 of the ICCPR.

The Human Rights Committee in the case of *Guerrero v. Colombia* considered the issue of the use of force by law enforcement officials, where the government had a suspicion that a guerrilla organization had kidnapped a former ambassador and was holding him hostage at a house. When the house was visited no hostage was found, but the government forces nevertheless waited for the return of the rebels and shot each of them at point-blank range, even though they were not armed at that time. The Committee concluded: "The police action was apparently taken without warning the victims and without giving them any opportunity to surrender to the police patrol or to offer any explanation of their presence or intentions. There is no evidence that the action of the police was necessary in their own defence or that of others, or that it was necessary to effect the arrest or prevent the escape of the persons concerned."[81] We can expect similar violations of human rights if autonomous weapons are used in such situations.

C. Dignity

The dignity of the human person is not only a fundamental right in itself, but constitutes the basis of fundamental rights in international law. Human dignity is inviolable and it must be respected and protected. The UDHR states in its preamble: "Recognition of the inherent dignity and of the equal and inalienable rights of all members of the human family is the foundation of freedom, justice and peace in the world." In ascribing inherent dignity to all human beings, the UDHR implies that everyone has worth that deserves respect.[82] There is an inextricable link between dignity and human rights.[83] The Vienna Declaration of the 1993 World Human Rights Conference affirms that all human rights derive from the dignity and worth inherent in the human person.

The use of lethal autonomous weapons could violate human dignity. An inanimate machine cannot respect the value of a human life or comprehend the significance of its loss. Allowing a machine to determine when to take away life would vitiate the importance attached to such decisions. According to Heyns, "There is widespread concern that allowing [LAWS] to kill people may denigrate the value of life itself."

D. Remedy and Reparation

Article 8 of the UDHR provides that everyone has the right to an effective remedy by the competent national tribunals for acts violating the fundamental rights granted him by the constitution or by law. Article 2(3) of the ICCPR requires states to ensure that any person whose rights or freedoms are violated shall have an effective remedy. A number of regional human rights treaties incorporate similar provisions.[84] The right to a remedy requires States to ensure individual accountability. It includes the duty to prosecute individuals for serious violations of human rights law. The 2005 Basic Principles and Guidelines on the Right to a Remedy and Reparation adopted by the UN General Assembly, speak of the obligation to investigate and prosecute. They require States to punish individuals who are found guilty.[85] The duty to prosecute applies to acts committed in law enforcement situations or armed conflict. The Fourth Geneva Convention of 1949 and its 1977 AP I also oblige States to prosecute grave breaches, i.e., war crimes, such as intentionally targeting civilians or knowingly launching a disproportionate attack.[86] The right to a remedy is not limited to criminal prosecution. It encompasses reparations, which can include restitution, compensation, rehabilitation, satisfaction and guarantees of non-repetition.[87]

Autonomous weapon systems could deny the right to a remedy. There are serious doubts that any meaningful accountability could be established for the actions of such weapons. The United States, for example, grants defence contractors immunity for harm caused by their weapons, and victims in any jurisdiction would often lack the resources and access to courts needed to bring a civil suit. This accountability gap would impair the ability of the law to deter harmful acts, and victims would be left dissatisfied with no one being punished for the suffering they experienced.[88] Transparency and accountability in the use of LAWS is another major concern.[89] Little is known about the extent to which different States are developing these weapons. States should make a disclosure about the extent to which they plan to develop autonomous systems and for what purposes.

E. Right Against Torture

The ICCPR, in Article 7, recognizes the right to be free from torture and "cruel, inhuman, or degrading treatment or punishment".[90] The UN Human Rights Council has found this to be binding as customary international law. The Human Rights Council has stated, "A number of international, regional and domestic courts have held the prohibition of cruel, inhuman or degrading treatment or punishment to be customary international law"….and "the prohibition of torture has been recognized as a "peremptory norm of international law". This makes the prohibition of cruel or inhuman punishment binding on all the states.

Article 9 of the ICCPR guarantees the right to physical security by stating, "Everyone has the right to liberty and security of person….No one shall be deprived of his liberty except on such grounds and in accordance with such procedure as are established by law." While this rule specifically envisions unlawful imprisonment, its scope could be much larger. For instance, by the use of lethal autonomous weapons government forces may control the movement of civilians. This would amount to an infringement upon their personal security and liberty because the victims would not be able to engage in lawful activities for fear of harm from the autonomous weapons.

The Rome Statute of the International Criminal Court defines torture as the intentional infliction of severe pain or suffering, whether physical or mental, upon a person in the custody of or under the control of the accused. Under this definition, even severe mental pain or suffering can constitute torture. The prolonged use of LAWS to hunt and kill suspected individuals could have a serious effect on a population. For example, if the civilians in an area are repeatedly injured or traumatized by LAWS attacks directed at non-state actors, they might develop psychological problems such as post-traumatic stress disorder. This would constitute severe mental pain and suffering violative of human rights under the ICCPR.

Conclusion

LAWS represent a qualitative shift in military technology. It is expected that in future they would be able to identify and attack targets without direct human involvement. Giving machines the power to release violent force without meaningful human control would cross a fundamental moral line and may lead to serious violations of IHL and human rights law. It is doubtful that an autonomous system would ever meet the requirements of IHL, such as the distinction between civilians and combatants. According to Liu (2012), IHL in its current manifestation is insufficient to regulate the growing use of LAWS. He attributes two reasons for this; first, the permissive nature of IHL based on military necessity, and second, the structural inability of IHL to cope with the challenges raised by this novel means and method of armed conflict.[91] The ATT, which has entered into force on 24 December 2014, requires a signatory State to assess the potential that a serious violation of IHL or international human rights law could be facilitated or committed with the arms or items to be exported, and not authorize the export of the items in question when there is an overriding risk of these violations. The UN Register of Conventional Arms impose limits on the trade in armed drones and but have no control over autonomous military systems,[92] which can be converted into lethal ones.[93] If the military commander, programmer or manufacturer cannot be held responsible for violations by LAWS, then it is feared that the 'responsibility gap' would allow using such weapon systems with impunity. According to Heyns, if the nature of a weapon renders responsibility for its consequences impossible, its use should be considered unethical and unlawful.[94]

Endnotes

1   Just war theory divides into three parts: (i) *jus ad bellum* – the justice of resorting to war; (ii) *jus in bello* – just conduct in war; and (iii) *jus post bellum* – justice at the end of war. *Jus in bello* means justice in war, and has traditionally been concerned with the treatment of the enemy. The rules of *jus in bello* are: (i) weapons prohibited by international law must not be used; (ii) distinction between combatants and non-combatants must be followed and only combatants may be targeted; (iii) the armed forces must use proportional force, i.e. proportional to achieving the end; (iv) prisoners of war must not be harmed and must be treated humanely; (v) no unlawful methods of war are permitted; and (vi) the armed forces are not justified in breaking these rules in response to the enemy breaking them.

2   Roff Heather M., Lethal Autonomous Weapons and Jus Ad Bellum Proportionality, *Case W. Res. J. Int'l L.*, Vol. 47, No. 1, 2015, p. 37-52.

3   Hurka Thomas, *Proportionality and the Morality of War*, Journal of Philosophy and Public Affairs, Vol. 33, 2005, p. 34-36.

4   Alston Philip, Lethal Robotic Technologies: The Implications for Human Rights and IHL, *Journal of Law, Information & Science*, Vol. 21 (2), 2011/2012.

5   Roff Heather M., Lethal Autonomous Weapons and Jus Ad Bellum Proportionality, *Case W. Res. J. Int'l L.*, Vol. 47, No. 1, 2015, p. 42.

6   Roff Heather M., Lethal Autonomous Weapons and Jus Ad Bellum Proportionality, *Case W. Res. J. Int'l L.*, Vol. 47, No. 1, 2015, p. 50.

7   The principle of distinction prohibits all means and methods that cannot make a distinction between those who take part in hostilities, i.e., combatants, and those who do not and are therefore protected. The rule that only military objectives may be attacked is based on the principle that, while the aim of a conflict is to prevail politically, acts of violence for that purpose may only aim at overcoming the military forces of the enemy. Acts of violence against persons or objects of political, economic or psychological importance may sometimes be more efficient to overcome the enemy, but are never necessary, because every enemy can be overcome by weakening sufficiently its military forces. Once its military forces are neutralized, even the politically, psychologically or economically strongest enemy can no longer resist. Sassoli Marco, Legitimate Targets of Attack under IHL, available at: http://www.hpcrresearch.org/sites/default/files/publications/Session1.pdf, accessed 3 December 2015.

8   An object is a military objective if, by its nature, location, purpose or use, it contributes effectively to the military action of the enemy and if its partial or total destruction, capture or neutralization offers a definite military advantage in the circumstances ruling at the time. Any object that does not fall under the definition of a military objective is a civilian object and must not be attacked. Article 52(2), AP I.

9   The principle of distinction is simple, but difficulty may arise when a target is of dual use. For instance a power grid, bridges, radio transmission towers, railway lines, can be classified as being civilian in nature as well as possessing a military purpose.

10   Sharkey Noel E., The Evitability of Autonomous Robot Warfare, *International Review of the Red Cross*, Vol. 94, No. 886, Summer 2012, p. 787-799.

11   Vilmer Jean-Baptiste Jeangene, Terminator Ethics: Should We Ban "Killer Robots" *Ethics and International Affairs*, 23 March 2015.

12   Sparrow Robert, Twenty Seconds to Comply: Autonomous Weapon Systems and the Recognition of Surrender, *International Law Studies*, Vol. 91, 2015, p. 299-728.

13   Article 41 of AP I clearly states that a person who is recognized or who, in the circumstances, should be recognized to be *'hors de combat'* shall not be made the object of attack; if he clearly expresses an intention to surrender.... provided that ... he abstains from any hostile act and does not attempt to escape.

14   Object recognition and classification will be particularly difficult in military applications given that wars often take place in complex and chaotic

environments, in various lighting conditions, and with smoke and fog obstructing the views of combatants. Krishnan Armin. 2009. *Killer Robots: Legality and Ethicality of Autonomous Weapons*, USA: Ashgate, p. 98.

15 Sparrow Robert, Twenty Seconds to Comply: Autonomous Weapon Systems and the Recognition of Surrender, *International Law Studies*, Vol. 91, 2015, p. 299-728.

16 Anderson Kenneth and Matthew Waxman, Law and Ethics for Autonomous Weapon Systems: Why a Ban Won't Work and How the Laws of War Can, American University Washington College of Law, Research Paper No. 2013-11, Columbia Public Law Research Paper, p. 11.

17 Rule 14 of the ICRC study on customary law provides: "Launching an attack which may be expected to cause incidental loss of civilian life, injury to civilians, damage to civilian objects, or a combination thereof, which would be excessive in relation to the concrete and direct military advantage anticipated, is prohibited." Jean-Marie Henckaerts and Louise Doswald-Beck, 2005, *Customary International Humanitarian Law*, Cambridge University Press, p. 46.

18 Article 51, paragraph 5 (b) of AP I.

19 The principle of proportionality is codified in article 51 (5) (b) of AP I and repeated in article 57. The requirement that an attack needs to be proportionate means that harm to civilians and civilian objects must not be excessive relative to the expected military gain. More specifically, even if a weapon meets the test of distinction, any use of a weapon must also involve evaluation that sets the anticipated military advantage to be gained against the anticipated civilian harm. Article 57, AP I contains three concrete obligations: (i) to refrain from launching a potentially disproportionate attack, (ii) to cancel or suspend an attack if it becomes apparent that the objective is not a military, and (iii) when a choice is possible between several military objectives for obtaining a similar military advantage, the objective which is expected to cause the least danger to civilian lives/ objects should be selected.

20 Arkin Ronald C., 2009. *Governing Lethal Behaviour in Autonomous Robots*, New York: CRC Press, p. 186.

21 Q&A on Fully Autonomous Weapons, Human Rights Watch, 21 October 2003.

22 Sharkey Noel E., The Evitability of Autonomous Robot Warfare, *International Review of the Red Cross*, Vol. 94, No. 886, Summer 2012, p. 789.

23 Kastan Benjamin, Autonomous Weapons Systems: A Coming Legal "Singularity"? *Journal of Law, Technology and Policy*, Vol. 2013, No. 1, p. 45-82.

24 US Air Force Judge Advocate General's Department, Air Force Operations and the Law: A Guide for Air and Space Forces, 2002, p. 27.

25 *Mind the Gap: The Lack of Accountability for Killer Robots*, Human Rights Watch, April 2015, p. 8.

26 Article 85, para 3 (b) of the Additional Protocol I and Article 8 (b) (iv) of the Rome Statute of the International Criminal Court. Margalit Alon, The Duty to Investigate Civilian Casualties During Armed Conflict, *Yearbook of International Humanitarian Law*, Vol. 15, 2012, Springer: Cambridge University Press, p. 155-186.

27 Anderson Kenneth and Matthew Waxman, Law and Ethics for Autonomous Weapon Systems: Why a Ban Won't Work and How the Laws of War Can, American University Washington College of Law, Research Paper No. 2013-11, Columbia Public Law Research Paper, p. 12-13.

28 The 1977 AP I, Article 57 2 (a), (i) and (ii) dealing with "Precautions in Attack" provides that those who plan or decide upon an attack shall: (i) Do everything feasible to verify that the objectives to be attacked are neither civilians nor civilian objects and are not subject to special protection but are military objectives.....and that it is not prohibited by the provisions of AP I; and (ii) Take all feasible precautions in the choice of means and methods of attack with a view to avoiding, and in any event to minimizing, incidental loss of civilian life, injury to civilians and damage to civilian objects.

29 Henderson Ian. 2009. *The Contemporary Law of Targeting: Military Objectives, Proportionality and Precautions in Attack under Additional Protocol I*, Leiden: Martinus Nijhoff Publishers, p.162.

30 The term "Feasible Precautions" has been defined under the 1980 CCW Protocol III (Prohibitions or Restrictions on the Use of Incendiary Weapons) as, "those precautions which are practicable or practically possible taking into account all circumstances ruling at the time, including humanitarian and military considerations.

31 Sandoz, Yves, Christophe Swinnarski and Bruno Zimmermann. 1987. *Commentary on Additional Protocols*. Geneva: ICRC, Martinus Nijhoff Publishers, p. 680.

32 Large teams of people and organizations are involved at all stages of the development of modern weapons, with extensive interaction between military decision makers, producers and those responsible for testing and approval for production. McFarland Tim and Tim McCormack, Mind the Gap: Can Developers of Autonomous Weapons Systems be Liable for War Crimes? Vol. 90, *International Law Studies*, 2014, p. 361-385.

33   It is a principle of international law, and even a general conception of the law, that any breach of an engagement involves an obligation to make reparation. PCIJ, *Chorzow Factory Case*, 1928, paragraph 103.

34   UN Doc. A/RES/60/47, Basic Principles and Guidelines on the Right to a Remedy and Reparation for Victims of Gross Violations of International Human Rights Law and Serious Violations of IHL, 21 March 2006.

35   State practice establishes this rule as a norm of customary international law applicable in both international and non-international armed conflicts. Rule 150, the ICRC customary Rules of IHL.

36   Under 2008 Cluster Munitions Convention "victim" means all persons who have been killed or suffered physical or psychological injury, economic loss, social marginalization or substantial impairment of the realization of their rights caused by the use of cluster munitions. They include those persons directly impacted by cluster munitions as well as their affected families and communities. Article 5 of the Convention dealing with "victim assistance" provides: "Each State Party with respect to cluster munitions victims in areas under its jurisdiction or control shall, in accordance with applicable IHL and human rights law, adequately provide age- and gender-sensitive assistance, including medical care, rehabilitation and psychological support, as well as provide for their social and economic inclusion. Each State Party shall make every effort to collect reliable relevant data with respect to cluster munitions victims.

37   Boven Theo van, 'Victims' Right to a Remedy and Reparations: The New United Nations Principles and Guidelines', in Ferstman Carla (ed.). 2009. *Reparations for Victims of Genocide, War Crimes and Crimes Against Humanity*, Leiden: Martinus Nijhoff Publishers, p. 19-40.

38   When violations of international law occur, perpetrators should not be allowed to escape accountability. Impunity for international crimes is a betrayal of our human solidarity with the victims of conflicts to whom we owe a duty of justice, remembrance, and compensation. For more details see: *Mind the Gap: The Lack of Accountability for Killer Robots*, Human Rights Watch, April 2015.

39   When it comes to a new weapon, it seems easy to get caught up in its new capabilities and ignore the larger issues concerning the morality or legality of its deployment, especially when the weapon's deployment likely means a decreased mortality rate for the deploying state's troops. Fry James, D., The XM25 Individual Semi-automatic Airburst Weapon System and International Law, *UNSW Law Journal*, Vol. 36, No. 2, p. 682-710.

40   Article 91of the 1977 AP I states: A Party to the conflict which violates the provisions of the Geneva Conventions or of this Protocol shall, if the case demands, be liable to pay compensation. It shall be responsible for all acts

committed by persons forming part of its armed forces.

41   In the last 15 years there have been numerous instances where drones have killed a large number of innocent civilians, including women and children in countries where there is no ongoing international armed conflict. In these countries compensation for the deaths and injuries has been an amount of US$ 3,500 (approximately) and 1,200 respectively. Civilian Harm and Conflict in Northwest Pakistan, Centre for Civilians in Conflict, (2010), p. 49-51, available at: http://civiliansinconflict.org/resources/pub/civilian-harm-and-conflict-in-northwest-pakistan, accessed 25 March 2016.

42   *Losing Humanity: The Case Against Killer Robots*, Human Rights Watch, November 2012, p.44.

43   Foy James, Autonomous Weapons Systems: Taking the Human out of IHL, *Dalhousie Journal of Legal Studies*, Vol. 23, 2014, p. 58.

44   Sullins John, When is a Robot a Moral Agent? *International Journal of Information Ethics*, Vol. 6, (2006) p. 23.

45   Sassoli Marco, Autonomous Weapons – Potential advantages for the respect of international humanitarian law, 2 March 2015, available at: https://phap.org/system/files/article_pdf/Sassoli-AutonomousWeapons.pdf, accessed 10 July 2015.

46   Under the principle of command responsibility, a commander is responsible for a subordinate's crimes if there is: (i) a senior-subordinate relationship; (ii) actual or constructive notice of the impending crime; and (iii) failure to take measures to prevent it. Also see: Mettraux Guenael. 2009. *The Law of Command Responsibility*, Oxford: OUP, p. 138.

47   Sparrow R., Killer Robots, *Journal of Applied Philosophy*, Vol. 24, No. 1, 2007, p. 66-70.

48   Geiss Robin, The International-Law Dimension of Autonomous Weapons Systems, October 2015, International Policy Analysis, Germany, p. 20.

49   Sparrow R., Killer Robots, *Journal of Applied Philosophy*, Vol. 24, No. 1, 2007, p. 70.

50   Geiss Robin, The International-Law Dimension of Autonomous Weapons Systems, October 2015, International Policy Analysis, Germany, p. 21.

51   *Mind the Gap: The Lack of Accountability for Killer Robots*, Human Rights Watch, April 2015, p. 30.

52   The US Department of Defence Directive 3000.09, p. 13-14.

53 Bills Gwendelynn, Laws unto Themselves: Controlling the Development and Use of Lethal Autonomous Weapons Systems, *The George Washington Law Review,* December 2014, Vol. 83 No. 1, p. 176-208.

54 *Mind the Gap: The Lack of Accountability for Killer Robots*, Human Rights Watch, April 2015, p.18.

55 If the nature of a weapon, or other means of war fighting, is such that it is typically impossible to identify or hold individuals responsible for the casualties that it causes, then it is contrary to this important requirement of *jus in bello.* Sparrow Robert, Killer Robots, *Journal of Applied Philosophy*, Vol. 24, (2007), p. 67.

56 Drones, once considered the state-of-the-art weapon systems have sparked worldwide debate, outrage, and devastation. They have proven their incapability of precision targeting and general unreliability during their use in the last 15 years. For more details see: Jha U C, *Drone Wars: Legal, Ethical and Strategic Implications*, KW Publications, New Delhi, 2014.

57 The UN General Assembly resolution, 'Guidelines for Conventional Arms Transfers' of 18 October 1991.

58 The United Nations negotiating process that led to the Arms Trade Treaty began in 2006 and ended in April 2013, when the UN General Assembly adopted the text of the treaty. Bauer Sibylle, Beijer Paul and Mark Bromley, The Arms Trade Treaty: Challenges for the First Conference of State Parties, SIPRI Insight on Peace and Security, No. 2014/2, September 2014, p. 1-12.

59 The United States, which was among the 12 original sponsors of the General Assembly resolution to adopt the treaty, joined with Western arms exporters and other European countries, as well as most African, Latin, and Pacific states, in approving the treaty. Iran, North Korea, and Syria voted against the accord. China, India, and Russia were among the abstentions. The final vote count was: 156 in favour, 3 against and 22 abstentions. The treaty has entered into force on 24 December 2014 and as on 15 November 2015 has 78 state parties. For details visit: http://disarmament.un.org/treaties/t/att.

60 Article 2 (1), the Arms Trade Treaty, 2013: This Treaty shall apply to all conventional arms within the following categories: (a) Battle tanks; (b) Armoured combat vehicles; (c) Large-calibre artillery systems; (d) Combat aircraft; (e) Attack helicopters; (f) Warships; (g) Missiles and missile launchers; and (h) Small arms and light weapons.

61 Article 6 of the Arms Trade Treaty dealing with prohibitions can be seen at Appendix A, at page 209 of the book.

62 Article 7, paragraph 3 of the ATT provides that if, after conducting the

assessment and considering available mitigating measures, the exporting State Party determines that there is an overriding risk of any of the negative consequences stated in Article 7, paragraph 1, the exporting State Party shall not authorize the export.

63　The ATT does not apply to all conventional weapons. It covers 'all conventional arms' within eight categories of weapons listed in Article 2, paragraph 1: battle tanks; armoured combat vehicles; large-calibre artillery systems; combat aircraft; attack helicopters; warships; missiles and missile launchers; and small arms and light weapons.

64　Bolton Matthew, Future-proofing the Draft Arms Trade Treaty: A Policy Brief, available at: https://politicalminefields.files.wordpress.com/2013/03/futureproofing-the-draft-arms-trade-treaty-42.pdf, accessed 20 November 2015.

65　Brandes Marlitt, "All's Well That Ends Well" or "Much Ado About Nothing"?: A Commentary on the Arms Trade Treaty, *Goettingen Journal of International Law*, Vol. 5, No. 2 (2013), p. 399-429.

66　The International Code of Conduct for Private Security Providers (the Montreux Document) was drafted with the aim to clarify international standards for the private security industry, especially when it operates in complex environments, and improve security companies' oversight and accountability. The Code was opened for signature on 10 November 2010. For more details see: The International Code of Conduct for Private Security Service Providers, Geneva Academy of International Humanitarian Law and Human Rights, Academy Briefing No 4, August 2013, available at: http://www.geneva-academy.ch/docs/publications/briefing4_web_final.pdf.

67　Kytomaki Elli, The Arms Trade Treaty's Interaction with Other Related Agreements, Chatham House, The Royal Institute of International Affairs, International Security Department, February 2015.

68　In recent years, the use of unmanned systems in warfare and other situations of violence has increased exponentially, and States continue to invest significantly into increasing the operational autonomy of such systems. These systems are used to perform a wide variety of tasks, most of which are unarmed. This may include, observation and reconnaissance, detection and neutralization of mines, improvised explosive devices (IED) and hazardous substances, wireless communications relay, as well as removal of obstacles and transport of supplies. These autonomous have also been weaponized for use in remote-controlled, semi-autonomous and full-automatic mode. Rapid development and proliferation of autonomous technology and the perceived lack of transparency and accountability of current policies have the potential of polarizing the international community, undermining the rule of law and, ultimately, of destabilizing the international security environment as a whole. Melzer Nils, Human Rights Implications of the Usages of Drones and Unmanned Robots in Warfare, Study Report for the European Parliament's Subcommittee on Human Rights, 2013.

69 Israel, while guarding its borders with the Gaza Strip, uses not only stationary sentry robots similar to the SGR-A1, but also the "Guardium", a remotely operated robotic vehicle which can be armed with lethal and non-lethal weapon systems. The Guardium is semi-autonomous in the sense that it is designed to perform routine missions, such as programmed patrols along border routes, but also to autonomously react to unscheduled events, in line with a set of guidelines specifically programmed for the site characteristics and security doctrine.

70 The Amnesty International has identified five key human rights issues for consideration in the current debate on LAWS. These are (i)) The scope of the Convention on Certain Conventional Weapons (CCW) does not cover non-conflict situations; (ii) L AWS will not be able to comply with relevant international human rights law (IHRL) and policing standards; (iii) Developments in existing semi-autonomous weapons technology pose fundamental challenges for the IHRL framework; (iv) In the absence of a prohibition, LAWS must be subject to independent weapons reviews; and (v) LAWS will erode accountability mechanisms. For details see: Amnesty International April 2015, Index: ACT 30/1401/2015.

71 The International Court of Justice has addressed the issue in an advisory opinion and found otherwise: The protection of the International Covenant on Civil and Political Rights does not cease in times of war, except by operation of Article 4 of the Covenant whereby certain provisions may be derogated from in a time of national emergency. Respect for the right to life is not, however, such a provision. In principle, the right not arbitrarily to be deprived of one's life applies also in hostilities. *Legality of Threat or Use of Nuclear Weapons, Advisory Opinion*, International Court of Justice, 8 July 1996.

72 Human rights law complements rules of IHL on the use of force during armed conflict, subject to a number of qualifications, including the proviso that during armed conflict the rules of human rights law are determined with reference to the legal rules of IHL. Further, people on both sides of the conflict retain their human rights such as the right to life and the right to dignity during armed conflict, even if the contents of the rights may differ according to the context. Christof Heyns, 'Autonomous weapons systems and human rights law', presentation made at the informal expert meeting organized by the state parties to the Convention on Certain Conventional Weapons, 13 – 16 May 2014, Geneva, Switzerland.

73 Universal Declaration of Human Rights (UDHR) was adopted by the UN

General Assembly on 10 December 1948. The General Assembly, while adopting the UDHR, proclaimed that it is a common standard of achievement for all people of all nations. The importance of UDHR can hardly be exaggerated; it has influenced the subsequent constitutional developments in most countries of the world, including India. It has triggered the evolution of a host of international instruments of human rights, both global as well as regional.

74 African Charter on Human and Peoples' Rights, adopted 21 June 1981 and entered into force 21 October 1986: Article 4 provides: "Human beings are inviolable. Every human being shall be entitled to respect for his life and the integrity of his person. No one may be arbitrarily deprived of this right"; American Convention on Human Rights adopted 22 November 1969, entered into force 18 July 1978, Article 4 provides: "Every person has the right to have his life respected. This right shall be protected by law and, in general, from the moment of conception. No one shall be arbitrarily deprived of his life"; European Convention for the Protection of Human Rights and Fundamental Freedoms, entered into force 3 September 1953, as amended by Protocols No. 3, 5, 8, and 11, Article 2 provides: "Everyone's right to life shall be protected by law. No one shall be deprived of his life intentionally save in the execution of a sentence of a court following his conviction of a crime for which this penalty is provided by law."

75 The 1979 Code of Conduct for Law Enforcement Officials was adopted by UN General Assembly Resolution 34/169 on 17 December 1979. The 1990 Basic Principles on the Use of Force and Firearms by Law Enforcement Officials was adopted by the Eighth United Nations Congress on the Prevention of Crime and the Treatment of Offenders, Havana, Cuba, from 27 August to 7 September 1990.

76 Article 2 and 3, the 1979 Code of Conduct for Law Enforcement Officials.

77 General Provisions, Principle 3 and 4, the 1990 Basic Principles on the Use of Force and Firearms by Law Enforcement Officials.

78 General Provisions, Principle 5, the 1990 Basic Principles on the Use of Force and Firearms by Law Enforcement Officials.

79 Christof Heyns, Autonomous Weapons Systems and Human Rights Law, Presentation made at the informal expert meeting organized by the state parties to the Convention on Certain Conventional Weapons 13 – 16 May 2014, Geneva, Switzerland.

80 The ICCPR, Article 14 provides: In the determination of any criminal charge against him, or of his rights and obligations in a suit at law, everyone shall be entitled to a fair and public hearing by a competent, independent and impartial tribunal established by law. In the determination of any criminal

charge against him, everyone shall be entitled to the following minimum guarantees, in full equality:(a) To be informed promptly and in detail in a language which he understands of the nature and cause of the charge against him; (b) To have adequate time and facilities for the preparation of his defence and to communicate with counsel of his own choosing; (c) To be tried without undue delay; (d) To be tried in his presence, and to defend himself in person or through legal assistance of his own choosing; (e) To examine the witnesses against him; (f) To have the free assistance of an interpreter if he cannot understand or speak the language used in court; and (g) Not to be compelled to testify against himself or to confess guilt. Everyone convicted of a crime shall have the right to his conviction and sentence being reviewed by a higher tribunal according to law.

81 The Human Rights Committee, Communication No. R. 11/45, Views, 31 March 1992, UN Doc. Supp. No. 40(A/37/40).

82 Jack Donnelly, 'Human Dignity and Human Rights', in Swiss Initiative to Commemorate the 60th Anniversary of the UDHR, Protecting Dignity: Agenda for Human Rights, June 2009, available at: http://www.udhr60.ch/report/donnelly-HumanDignity_0609.pdf, accessed 10 October 2015.

83 The Preamble to the ICCPR provides "….recognition of the inherent dignity and of the equal nd inalienable rights of all members of the human family is the foundation of freedom, justice and peace in the world;. ….these rights derive from the inherent dignity of the human person."

84 Article 7 (1) of the African Charter on Human and People's Rights provides: Every individual shall have the right to have his cause heard. This comprises: (a) the right to an appeal to competent national organs against acts of violating his fundamental rights as recognized and guaranteed by conventions, laws, regulations and customs in force. Article 25 of the American Convention on Human Rights provides: Everyone has the right to simple and prompt recourse, or any other effective recourse, to a competent court or tribunal for protection against acts that violate his fundamental rights recognized by the constitution or laws of the state concerned or by this Convention. Article 13 of the European Convention for the Protection of Human Rights and Fundamental Freedoms, states: Everyone whose rights and freedoms as set forth in this Convention are violated shall have an effective remedy before a national authority.

85 Article 4, Basic Principles and Guidelines on the Right to a Remedy and Reparation for Victims of Gross Violations of International Human Rights Law and Serious Violations of International Humanitarian Law; UN General Assembly, Resolution 60/47, 16 December 2005.

86 Article 146, The Fourth Geneva Convention Relative to the Protection of

Civilian Persons in Time of War, 12 August 12, 1949; and Article 86, the 1977 Additional Protocol I.

87 Articles 18-23, the 2005 Basic Principles and Guidelines (UN Doc. A/RES/60/47).

88 Solomon Cara, The Human Rights Implications of Killer Robots, 16 June 2014, available at: http://hrp.law.harvard.edu/staff/the-human-rights-implications-of-killer-robots/, accessed 6 July 2015.

89 For more details see: *Shaking the Foundations: The Human Rights Implications of Killer Robots*, Human Rights Watch Report, 2014, p. 33.

90 The European Commission on Human Rights has clarified the difference between torture, cruel treatment, inhuman treatment, and degrading treatment. It determined that the difference was largely a matter of degree. Torture subsumes cruel, inhuman, and degrading treatment, and it is typically used to obtain information or a confession but can also be used as punishment. Inhuman treatment is unjustifiably inflicting severe physical or mental suffering while degrading treatment is grossly humiliating a person or driving one to act contrary to one's will or conscience. Cruel treatment is an unjustifiable infliction of physical or mental suffering.

91 Liu Hin-Yan, Categorization and legality of autonomous and remote weapons systems, *International Review of the Red Cross*, Vol. 94, No. 886, Summer 2012, p. 627-652.

92 The current arms control tools are not suited to the age of robotics, an age that is no longer in the realm of science fiction. Singh Harjeet, Robotics and the Changing Characteristics of Warfare, *CLAWS Journal*, Winter 2014, p. 92.

93 In 2014, it was reported that the Swiss armed forces are likely to purchase Hermes-900 type reconnaissance drones from Israel as part of its 2015 arms procurement programme. These drones could be easily further developed into lethal autonomous weapons.

94 Heyns C., *Report of the Special Rapporteur on extrajudicial, summary or arbitrary executions*, UN General Assembly Doc. A/HRC/23/47, 9 April 2013, paragraph 80.

V Ethical and Moral Issues

Introduction

An autonomous weapon system, once activated, can select and engage targets with lethal force without further intervention by a human operator. A few such systems, which human operators can activate and the operation of which can be overridden by human beings, already exist. These include Israel's Iron Dome anti-missile system and the American Patriot and Phalanx antimissile systems. The South Korean lethal sentry SGR-A1, which has the capability to shoot without human intervention, can be used against hostile intruders at international borders. The future of such weapon systems appears bright as at least 40 countries are trying to acquire or develop them for their militaries. The US government and military anticipate the automation of armed conflict by 2032. These technological developments in weaponry raise serious questions relating to IHL, human rights, and disarmament agreements. Experts are of the strong opinion that the spread of LAWS might lead to a new global arms race, as well as to the lowering of the threshold for entering into an armed conflict.[1]

The autonomous military systems have a valuable role to play in future armed conflict and no sagacious person can deny this. These systems, which are capable of surveillance, neutralization of IEDs, clearing minefields, etc., will be deployed in greater number in future conflicts. All the military autonomous systems currently in use have a person in the loop to control their flight and apply lethal force. However, with advancement in technology, the scenario is likely to change, with the autonomous system being given all the capabilities, including even the use of lethal force over a person. The US has made it clear that it has the intention to develop and use autonomous weapon systems for aerial, ground and underwater operations. The US is not the only country with this aim—China, Russia, Israel, the UK and a few other States may follow suit. The aim would be to develop and deploy a network of land, sea and aerial LAWS that would operate autonomously to locate their targets and destroy them without human intervention. The development and use of LAWS raises many ethical issues.

LAWS is not the first weapon that has fomented the issue of weapon ethics. The use of military drones by the US and the UK for targeted killing in Pakistan, Afghanistan, Somalia and Libya also caused anguish amongst legal experts and scholars. The main concern was the manner in which the American CIA used the weapon systems to undertake extra-judicial killings in countries which were not at war with the US. The unique features of LAWS, i.e. that it allows the killing of adversaries at a distance and that the weapon can take a decision on the use of lethal force against a human being, single it out for ethical examination.

Though law and ethics are very different, a number of persons conflate the two, asserting that everything that is legal must be ethical, and everything that is illegal must be unethical. While law and ethics are closely related, they are not the same.[2] The questions raised by these two disciplines are different.[3] The term 'ethics'[4] broadly includes not just normative issues, i.e. questions about what we should or ought to do, but also general concerns related to social and cultural impact, as well as risk arising from the use of LAWS, e.g. responsibility after a malfunction.

The four Geneva Conventions of 1949, the 1977 Additional Protocols, weapon ban/regulation treaties and the just war theory contain a set of ethical principles. These have been codified into international agreements and are referred to as the laws of armed conflict or IHL. The fundamental principles of humanity and military necessity form the basis for determining the balance between military requirements and the need to limit the suffering and devastation caused by an armed conflict, and provide protection to those most vulnerable, such as the wounded and captured, and the civilian population. These two main principles are complemented by several other fundamental principles which are drawn from the two main ones. Together, they form the overall system. The other fundamental principles include the principle of distinction, the principle of proportionality, and the principle of prohibition of unnecessary suffering or superfluous injury in the use of certain types of weapons and means of combat. These principles are interrelated and form a system, which have been recognized throughout long history.

Ethical Implications

Autonomous systems or robots have improved quality and productivity in the industrial sector. They are driving cars and there are fully autonomous cars in the pipeline. In medical science, robots are being used for complicated surgical procedures. At home, human-like robots and robotic pets are being considered friends and companions. They can play musical instruments, sing, dance, etc. to please and comfort human beings. A wide variety of robots is now being manufactured and there is a strong possibility that human beings will coexist with robots in the near future. The autonomous systems combine many scientific disciplines: mechanics, automation, electronics, computer science, cybernetics, artificial intelligence and information technology. They also draw elements from several other fields of knowledge, such as biology, physiology, neuroscience, as well as humanistic disciplines like linguistics and psychology. The robotic revolution in military weaponry, however, raises a number of ethical, moral and legal questions.[5]

Isaac Asimov was the first to put forward ethical principles for robotics in a science fiction work, in which he formulated his Three Laws of Robotics.[6] First, "A robot may not injure a human being, or through inaction, allow a human being to come to harm". Second, "A robot must obey the orders given it by human beings, except where such orders would conflict with the First Law". Third, "A robot must protect its own existence as long as such protection does not conflict with the First or Second Law". Asimov later introduced a fourth or zeroth law that outranked the others: "A robot may not harm humanity, or, by inaction, allow humanity to come to harm". Since then, Asimov's laws of robotics have become a key part of science fiction culture. Despite not being directly applicable to the real world, Asimov's laws are currently used in various countries as a starting point for drawing up roboethical guidelines.[7] In recent years, roboticists have made rapid advances in the technologies that are bringing the kind of advanced robots envisaged by Asimov closer to us. Robotic scientists believe that regulating robots' behaviour is going to become more difficult in the future, since they will increasingly have self-learning mechanisms built into them. It may become impossible to predict the behaviour of robots fully, since they will not be behaving in predefined ways but will learn new behaviour as they grow.[8]

In August 2015, in Gurgaon, an industrial hub about 20 km from Delhi, sharp welding sticks jutting out of the robotic arm of a machine pierced Ramji Lal, a 24-year-old worker, killing him on the spot. The worker had apparently moved too close to the robot while adjusting a metal sheet that had come unstuck. This was the first robotic killing in India. In 1981, Kenji Urada, a 37-year-old Japanese factory worker, climbed over a safety fence at a Kawasaki plant to carry out some maintenance work on a robot. In his haste, he failed to switch the robot off properly. Unable to sense him, the robot's powerful hydraulic arm kept on working and accidently pushed the engineer into a grinding machine. This was the first recorded death at the hands of a robot. In the last 25 years, there have been a number of gruesome industrial accidents. People have been crushed, hit on the head, welded and even had molten aluminium poured over them by robots. Even though military robots (or LAWS) may be perceived of as weapons that will give future generations clear benefits on the battlefield, serious risks may arise when it comes to machines which are capable of lethal action and which will acquire more autonomy. Hence, there is an urgent need to evaluate the consequences of new robotic weapon systems of warfare.

Lin (2008) has emphasized three fundamental ethical implications of autonomous weapon systems: (i) whether LAWS would be able to follow established guidelines of IHL and the rules of engagement, as specified in the Geneva Conventions; (ii) whether they would know the difference between military and civilian personnel; and (iii) whether they would recognize a wounded soldier and refrain from shooting. [9] Sparrow (2011) has highlighted a number of ethical issues relating to the development and deployment of autonomous weapons in warfare. [10] While most of these issues are unique to LAWS, some already apply to the existing weapons, including drones. These issues are: distant killings, discrimination and proportionality, violations of the rules of targeting, undermining of 'warrior virtues', and immorality. According to Johnson (2014), the important ethical concerns raised by the use of LAWS in armed conflict are those pertaining to responsibility and accountability. Who will be responsible when robots decide for themselves and behave in unpredictable ways or in ways that their human partners do not understand? For example, who will be responsible if an autonomously operating unmanned aircraft crosses a border without authorization or erroneously identifies a friendly aircraft as a target and shoots it down?[11] Robin (2015) has analysed the ethical aspects of LAWS by posing three basic questions: (a) whether the danger of armed conflicts would increase due to the deployment of autonomous weapons systems; (b) whether LAWS could comply with the rules of IHL, i.e. distinction and proportionality; and (c) whether the deployment of LAWS would violate human dignity.[12]

Lowering Threshold of Armed Conflicts

It has been advocated by a number of military analysts that the threshold for the deployment of military force would be lowered with the increased availability of LAWS. According to Asrao, the development of robotic weapons will make it easier for political leaders to take an unwilling nation to war.[13] In support of his argument, he cited the examples of the 1991 Persian Gulf War, the 1999 war in Kosovo and the 2003 invasion of Iraq. Another example is the ongoing armed conflict in Syria, where a number of States like the US, the UK and Russia have deployed their air forces and navies for action against the ISIS. According to Asrao, these events have brought into sharp relief the complex relationships between the political requirements of national leaders, the imagery and rhetoric of propaganda and the mass media, and the general will of citizens in the processes of deciding when a democratic nation will go to war.[14]

The main concern of political and military leaders in an armed conflict remains the lives of its citizens as well as of the combatants. A political strategy evolved in response to this is to resort to relatively 'safe' forms of fighting in order to limit casualties, and to invest in technologies that promise to lower the risks and increase the lethal effectiveness of their military. This was evident in the NATO involvement in the 1999 Kosovo conflict, in which NATO limited its military operations to air strikes. The use of technologically advanced weapons such as LAWS can shift risks away from a nation's own soldiers, thus prompting a State to initiate an armed conflict and use such weapons to kill enemy soldiers and win.

Lethal autonomous weapons would remove the combatants who operate them from the area of armed conflict and reduce the risk of causalities for the State/ non-State actors who possess them. They would also reduce the political costs and risks of engaging in armed conflict. This may result in an overall lowering of the threshold of initiating an armed conflict. LAWS also have the potential to cause regional or global instability and insecurity and fuel arms races. Once owned by non-State actors, they would result in the escalation of conflicts at unexpected geographical locations. Therefore, LAWS pose a serious threat to international stability and the ability of international bodies to manage armed conflicts.[15] Once LAWS are available to the armed forces, the latter would like to test and use them for various contingencies and operations. The presence of LAWS could thus prompt States to use such weapons during internal armed conflicts and border skirmishes, and for securing facilities and territory,[16] domestic policing, crowd control, guarding prisoners, and other security applications involving the use of force.

Distant Killing

The next key issue which arises in any discussion of LAWS is the ethics of killing at a distance.[17] The gradual increase in the sophistication of military machinery has had the effect of moving soldiers farther and farther away from their enemy.[18] The pilots who operated lethal drones in the US were sitting thousands of miles away from the people they killed. The drones they operated were flying so high and so quietly that the people they targeted had no idea till the missiles arrived. Most of the LAWS being developed around the world would have the capability of detecting and firing at targets while maintaining a very safe distance. The robot sentry deployed by South Korea in the demilitarized zone uses sensing equipment to detect human beings as far as two miles away as it patrols the frontier, and can kill them from a safe distance.

Killing from a distance, that too, by a machine, raises serious ethical concerns. If the person who pushes the button does not see what he is doing to those on the ground, it is questionable whether the killing is justified. Killing from a distance is generally easier than killing from close by. Distance makes killing too easy as the natural moral–psychological barrier to killing is removed. In distant killing by LAWS, there is no place for empathy; it is worse than a computer game, in which an attacker may have some remorse. Military commanders and political leaders have to seriously consider whether the use of LAWS is at all compatible with a respect for the humanity of those being killed.

Warrior Virtues and LAWS

Military values and virtues have a unique importance both for the military profession and civilian society. The citizens of a country hold armed forces personnel in high esteem by virtue of a strong moral and ethical value system. Military training lays great emphasis on the development of physical courage, fortitude, loyalty, willingness to sacrifice for the sake of the community, and even love and compassion. The fact that there is a correlation between war and virtue plays an important role in the culture of military organizations. It is difficult to visualize how a combatant is going to demonstrate his military virtues while LAWS are in battlefield. Keeping human soldiers out of combat by allowing machines to fight will deprive a soldier of the opportunity to exercise the virtues currently considered characteristic of a good warrior. How will a soldier demonstrate his courage and compassion when LAWS are being used to eliminate enemy combatants? No machine can ever replace the key elements of military virtues that make human beings irreplaceable in making lethal decisions on the battlefield. Military personnel who are to effectively serve society must have certain character traits, such as honesty, selflessness and commitment, which inspire trust and confidence.[19] What would happen to these traits if a combat zone were occupied by LAWS, which had complete autonomy to use lethal force? The development and use of LAWS would also pose a serious threat to the existing warrior virtues in the armed forces.

Ethical LAWS

According to Professor Ronald Arkin (2009), ethical robots have four components: (1) an ethical governor—a series of algorithms which determine whether a lethal response is ethical, on the basis of the rules of IHL that constrain lethal action; (2) an ethical behavioural control— the capacity to produce lethal responses that fall within acceptable ethical bounds; (3) an ethical adaptor—the ability to update the robot's constraints and ethically related behavioural parameters; and (4) a responsibility advisor—a part of the human–robot interaction component used for permission planning and managing overrides by the operator. The responsibility advisor gives advice in advance of the mission to the operator/ commander on his ethical responsibilities should LAWS be deployed for a specific battlefield situation.[20] According to Arkin, some of the critical issues relating to an ethical LAWS are: the use of proactive tactics or intelligence to enhance target discrimination; recognition of a previously identified legitimate target as surrendered or wounded (a change to POW status); discrimination in battlefield conditions; proportionality and the use of a given set of weapon systems and methods of employment; and the battlefield assessment of military necessity. These issues are based on formalized logical statements of IHL and mission-specific ROE, which can be set by a commander before deploying LAWS in an armed conflict.[21] However, this translation of legal principles and rules into an algorithm-compatible rule has yet to be achieved. Arkin concludes, "It is not my belief that LAWS will be able to be perfectly ethical in the battlefield, but….they can perform more ethically than human soldiers are capable of." Substantial additional basic and applied research must be conducted before any such system can even be considered for use in an armed conflict.[22] However, Sparrow is of the firm belief that any use of LAWS is unethical due to the *jus in bello* requirement that someone must be responsible for a possible war crime. To support his argument, Sparrow draws a parallel between LAWS and child soldiers, both of which he claims cannot assume moral responsibility for their action.[23]

The following were the findings of a survey conducted on the ethical aspects of the use of LAWS in an armed conflict, covering participants ranging from 19 to 66 years of age from all over the world, including Australia, Asia, eastern Europe and Africa.

- A human soldier was the most acceptable entity in warfare, followed by a robot as an extension of the war-fighter, with LAWS being the least acceptable.

- Taking of human life by LAWS both in open warfare and covert operations was unacceptable to more than half of the participants.

- More of the military and policy-makers were in favour of the same ethical standards for both soldiers and robots than the general public and roboticists, who preferred higher standards for robots.

- The most acceptable role for using both types of robots (as an extension of the war-fighter or LAWS) was reconnaissance.

- The majority of the participants agreed that the ethical standards, namely the IHL, ROE, code of conduct and moral standards, should apply both to soldiers and robots.

- Nearly 60% of the participants believed that a LAWS should have the right to refuse an order that it finds unethical.

- Sixty-six per cent of the participants were in favour of higher ethical standards for a robot than for a soldier.

- The majority of the participants believed that it would be easier to start wars if LAWS were employed in an armed conflict.[24]

In conclusion, the respondents felt that autonomous systems may have a place in warfare, but they should preferably not exhibit any force. Jakob Kellenberger, the former president of the ICRC, has been of the firm opinion that the deployment of LAWS raises a range of fundamental legal, ethical and societal issues which need to be considered before such systems are developed or deployed.[25] Ethics operates in a complex domain with some ill-defined legal moves. We have a limited understanding of what a proper ethical theory is. Not only do people disagree on the subject, but individuals can also have conflicting ethical intuitions and beliefs. The burden of evaluating ethical issues related to new military technologies needs to be shared by military researchers and policy-makers, and also by the wider public. The issues raised by LAWS are too real and we need to address them at the earliest. We cannot give destructive power to a machine. There is a need to devise a mechanism for 'human impact assessment' of LAWS, analogous to the environmental impact assessments required for new construction projects.[26] In an era in which the rights of endangered species of plants and animals are becoming increasingly important,[27] we must be equally concerned about developing LAWS and programming them to kill human beings. An international ban on LAWS can be established on the principle that the authority to decide to use lethal force cannot be legitimately delegated to an automated process and must remain the responsibility of a human being who has the duty to make an informed decision before taking human lives.[28] Can we ever trust LAWS to take strategic decisions and make political leaders and military commanders redundant?

Morality of LAWS

Lethal autonomous weapon systems may be programmed for patrolling or surveillance of a designated area and taking a lethal decision to kill or destroy a target. However, the use of LAWS raises many potential moral issues, whether or not their algorithms can be discriminatory enough to avoid killing civilians or destroying civilian property. The existing autonomous military systems, like the Korean SGR-A1, can identify and shoot without human intervention.[29] The Israeli *Harpy*—a loitering munition—can detect radar signals and destroy its target. These autonomous systems have serious limitations. For instance, the SGR-A1 cannot distinguish between a combatant and non-combatant and *Harpy* cannot distinguish whether the radar is on an antiaircraft station or on the roof of a hospital. Even if we can develop LAWS with sensor technology and programmed algorithms capable of distinguishing between a civilian and a combatant, the moral issue of whether we can task a machine to go to the battlefield to kill human enemies remains.

In Hellstrom's view, assigning moral responsibility to LAWS is a possibility. The futuristic robots, according to him, will be equipped with highly advanced cognitive abilities to perceive, plan and learn, and also with a repertoire of complex behaviours and capabilities to operate lethal weapons of various kinds. With an increase in the autonomous power of these armed intelligent robots, military commanders could treat them as human soldiers. The robots would receive orders and ROE, and would be expected to follow them in the same way as human soldiers do. He, however, concludes that it would indeed be highly immoral to develop robots capable of performing acts involving life and death.[30]

LAWS engineered to make decisions on killing might become technically better at complying with IHL. They may calculate better than humans and be equipped with superior sensors. However, this ignores the cardinal consideration of moral agency, which is a defining characteristic of being human. Machines operate as programmed; they are incapable of making moral judgments. They are devoid of empathy, love, generosity and mercy. We must avoid a situation in which an enemy combatant stands in front of a machine, pleading to be taken prisoner but eventually being shot dead. We cannot absolve ourselves of responsibility if LAWS run amok. Whatever the advancement in technology or artificial intelligence, the use of lethal force must always require human judgment.

In an armed conflict, there are occasions when the circumstances of battle require brutality and the application of the maximum violence. Even during such battles, there have been ample examples of ruthless combatants showing mercy to enemy soldiers. Can we expect such morality from an

autonomous machine? Johnson and Axinn, while debating on "whether the decision to take a human life be relinquished to LAWS", have opined that such weapons must not be developed for the following reasons: (i) developing such weapons would amount to treating a human being as an object, instead of a person with inherent dignity;[31] (ii) the weapon can only mimic moral actions, and cannot be moral; (iii) the weapon would have no human emotions and no feelings about the seriousness of killing a human being; (iv) it would follow an illegal order, if programmed to do so; (v) it would not realize the enormity of the error of killing the 'wrong' person; and (vi) its use would be a violation of military honour. Johnson and Axinn conclude that the use of LAWS in armed conflict should be banned.[32]

As is clear by now, an autonomous weapon system, howsoever efficient, cannot completely replace the presence of a moral agent in the form of a human being. Though there have been a number of instances in which human beings have failed in their moral obligations, they can never be replaced with a machine which is programmed to take lethal decisions. The application of lethal violence can never be delegated to LAWS. In an armed conflict, a combatant is aware of the lethality of the weapons being used. He may get hurt or killed. However, machines cannot be 'hurt' or 'killed', though they are also vulnerable, albeit in a very different way, and they do not experience their vulnerability.[33] This implies that LAWS cannot know what they are doing to human beings. They cannot even understand the threat they are posing to human beings. Therefore, killing by machine must not be automated in any way; and if killing in an armed conflict is justified at all, human beings should always be associated with the decision to use lethal force.

Individuals and States in peacetime, as well as combatants, military organizations and States in armed conflict situations, have a duty not to delegate the authority or capability to initiate the use of lethal force to a machine. An international ban on LAWS can be established on the principle that the authority to use lethal force cannot be legitimately delegated to a machine. In the armed forces, there is a hierarchy from the highest military commander down to a soldier, and at every level, there is a responsible human being to bear both the authority and responsibility for the use of force. The principle of command responsibility does not allow one to abdicate one's moral and legal obligations to determine that the use of force is appropriate in a given situation.[34] It might be transferred to another

responsible human being, but one then has a duty to oversee the conduct of that subordinate. LAWS are not responsible human agents; one cannot delegate this authority to them. The responsibility to use lethal force must remain with a human being, who has the duty to make a decision before taking human lives.

Before a state decides to develop and use LAWS against its adversaries, it must answer a few questions. Can a robot be trusted with the delivery and care of a child? Can a robot be used for the evacuation and treatment of a battle casualty? Can a robot be entrusted with counselling a child or psychiatric patient? Can the state replace a 'judge' with a robot to decide a case in a just and fair manner? If the answer to any of these questions is in the negative, the state must not allow a machine to take a decision on killing a human being.

In the context of the use of LAWS, let us take up the example of nuclear weapons. Nuclear weapons were developed in the early 1940s and two of them were used immediately in 1945.[35] It is estimated that there were 140,000 casualties in Hiroshima and 75,000 in Nagasaki. Nuclear weapons also cause severe damage to the climate and environment on a scale incomparable to any other weapon. No sane person would agree that nuclear weapons could ever abide by the rules of IHL and international human rights law.[36] Efforts are being made to remove nuclear weapons from military arsenals altogether.[37] The moral implications of nuclear weapons were not publicly debated until after their use, and many of the scientists who were involved in their development later regretted ignoring those moral issues.

Conclusion

We must explore the moral issues related to LAWS before they are developed and deployed in an armed conflict. For the preservation of human morality, dignity, justice and law, we cannot accept the possibility of LAWS making the decision to take a human life.[38] It is human beings who must ultimately bear moral responsibility and face the horror of war. They must remain in control of a weapon and manage its behaviour. There is a need to impose a ban on the development and use of highly indiscriminate LAWS. If we fail, it would set a dangerous precedent for the development of even more indiscriminate and inhumane means and methods of warfare in the future.

Endnotes

1   Sparrow, Robert, Killer Robots, *Journal of Applied Philosophy*, Vol. 24, No. 1, 2007, p. 62–77.

2   Coleman Stephen, 'Ethical Challenges of New Military Technologies', in Nasu Hitoshi and Robert McLaughlin (ed.). 2014. *New Technologies and the Law of Armed Conflict*, The Hague: TMC Asser Press, p. 30.

3   While there is an intimate relationship between law and ethics, they are certainly not the same thing. For example, something may be: (a) legal but unethical, such as the apartheid laws of South Africa; (b) illegal but ethical, such as exceeding the speed limit to get a critically injured person to the hospital; or (c) ethical but not enforced by law, such as the tenet that parents ought to love their children. Coleman Stephen, Possible Ethical Problems with Military Use of Non-Lethal Weapons, *Case W. Res. J. Int'l L.*, Vol. 47 (1), (2015), p. 187.

4   There are strong contrary views on the 'ethics' of autonomous weapons: "Segments of the public involved in these discussions (new military technologies) harbour distinctive and incompatible—and sometimes conceptually confused and unclear—notions of what 'ethics' entail. From individual and culturally determined intuitions regarding morally right conduct, through the achievement of beneficial outcomes, all the way to equating ethics to mere legal compliance, this discord results in frequent and virtually hopeless equivocation." Lucas George R., Legal and Ethical Percepts Governing Emerging Military Technologies: Research and Use, *Utah Law Review*, 2013, No. 5, p. 1271–1281.

5   Gianmarco Veruggio, a member of the European Robotics Research Network (EURON), has coined the term 'roboethics' to refer to the ethics applied to robotics. He defined the objective of roboethics as the development of scientific, cultural and technical tools in order to promote the development of robotics for the advancement of human society and individuals, and to help prevent its misuse against humankind.

6   Isaac Asimov was born in 1919 into a Russian-Jewish family that immigrated to the US three years later. The Three Laws of Robotics (also known as Asimov's Laws) are a set of rules introduced by Asimov in his short story "Runaround" in 1942. Asimov attributes the three laws to John W. Campbell, from a conversation that took place on 23 December 1940. According to Asimov, tools created for everyday human convenience tend to feature safety elements. Electrical wiring is insulated, pressure cookers have relief valves, and so on. Therefore, robots should surely be built with safeguards aimed at preventing injury to humans.

7   Pau Mauro, Robots in warfare: Why we need to start an ethical debate,

available at: http://folk.uio.no/mauro/papers/robowars.pdf, accessed 7 July 2015.

8 Trust me, I'am a robot, *The Economist*, 8 June 2008; Professor Dr Robot QC, *The Economist*, 17 October 2015.

9 Lin Patrick, George Bekey and Keith Abney, Autonomous Military Robotics: Risk, Ethics, and Design, Report prepared for the US Department of Navy, Office of Naval Research, 2008, p. 21.

10 Sparrow, R., 'Robotic Weapons and the Future of War', in Jessica Wolfendale and Paolo Tripodi (eds), 2001, *New Wars and New Soldiers: Military Ethics in the Contemporary World*. Surrey, UK & Burlington, VA: Ashgate, p. 117-133.

11 Deborah G. Johnson and Noorman, Merel E., Responsibility Practices in Robotic Warfare, *Military Review*, May-June 2014, p. 12-21.

12 Geiss Robin, The International-Law Dimension of Autonomous Weapons Systems, October 2015, International Policy Analysis, Germany.

13 Asaro Peter M., How just could a robot war be? Available at: http://peterasaro. org/writing/Asaro%20Just%20Robot%20War.pdf, accessed 10 November 2015.

14 Asaro Peter M., How just could a robot war be? Available at: http://peterasaro. org/writing/Asaro%20Just%20Robot%20War.pdf, accessed 10 November 2015.

15 Asaro Peter, On banning autonomous weapon systems: human rights, automation, and the dehumanization of lethal decision-making, *International Review of the Red Cross*, Vol. 94, No. 886, Summer 2012, p. 687–709.

16 For instance, South Korea has deployed autonomous robots armed with machine guns and live ammunition along the border with North Korea. The system is designed to shoot at any human attempting to cross the demilitarized zone.

17 For thousands of years, combatants have continually moved farther and farther away from the point of direct physical engagement during battle. The longbow, musket, cannon and airplane are a few examples of technology that provide an extended reach to those who possess such instruments of warfare, and simultaneously reduce the risk of death should one retain the technological advantage. Campo, Joseph L., Distance in War: The experience of MQ-1 and MQ-9 aircrew, *Air & Space Power Journal*, 2015, p. 3–10.

18 IHL and the Challenges of Contemporary Armed Conflict, 32nd International Conference of the Red Cross and Red Crescent, Geneva: ICRC, 2015, p. 44.

19 Nagar Samrath, Morals and Ethics in the Armed Forces: A De Novo Look, *CLAWS Journal*, Summer 2009, p. 198–208.

20 Arkin Ronald C. 2009, *Governing Lethal Behaviour in Autonomous Robots*, USA: Chapman & Hall/CRC, p. 125.

21 For more details see: Arkin Ronald C. 2009, *Governing Lethal Behaviour in Autonomous Robots*, USA: Chapman & Hall/CRC, p. 125–153.

22 Arkin Ronald C. 2009, *Governing Lethal Behaviour in Autonomous Robots*, USA: Chapman & Hall/CRC, p. 98.

23 One thing that is unethical about employing youth below a certain age in combat duties is that life and death decisions are placed in the hands of children who cannot be held responsible for their actions. While they lack full moral autonomy—and, therefore, are not morally responsible for what they do—there is clearly a sense in which children are autonomous. They are capable of a wide range of decisions and actions. They are certainly much more autonomous than any existing robot. Yet they are not appropriate objects of punishment, as they are not capable of understanding the full moral dimensions of what they do and, therefore, of understanding the connection between their punishment and their crime. Sparrow, R., Killer Robot, *Journal of Applied Philosophy*, Vol. 24(1), March 2007, p. 62–77.

24 The survey was conducted online by a commercial survey company <*SurveyMonkey.com*>, and the use of human subjects was approved using formal institutional review board procedures. It followed the recommendations and guidelines espoused for the preparation and conduct of internet surveys. The full survey structure, design and results are reported in detail in a lengthy technical report See: Moshkina, L. and Arkin, R., Lethality and Autonomous Systems: Survey Design and Results, Technical Report, 2008, The US Army Research Office.

25 Jakob Kellenberger, Keynote Address, IHL and New Weapon Technologies, 34th Round Table on Current Issues of IHL, San Remo, Italy, 8–10 September 2011, p. 5–6.

26 Singer, P. W. 2009. *Wired for War: The Robotics Revolution and Conflict in the 21st Century*, New York: Penguin books, p. 427.

27 The Government of India cleared a proposal in 2014 for setting up a radar station for the armed forces at Narcondam Island. This tiny island is part of the Andaman and Nicobar Island group in the Bay of Bengal, and the only home of the endemic Narcondam hornbill. The existing police outpost at the island has itself caused some loss of habitat. It is dependent on the single freshwater source available on the island, and any additional human presence and installations will only compound the problems of the island and its unique

flora and fauna. In February 2012, the Director, Bombay Natural History Society, made a site visit on behalf of the Standing Committee of the National Board for Wildlife, a high-level statutory advisory body of the government, and strongly recommended the rejection of the proposal, as it poses a grave threat to the world's only population of Narcondam hornbills.

28 Asaro Peter, On banning autonomous weapon systems: human rights, automation, and the dehumanization of lethal decision-making, *International Review of the Red Cross*, Vol. 94, No. 886, Summer 2012, pp. 687–709.

29 The SGR-A1 Security Robot is a stationary system developed for the Korean Demilitarized Zone. It is able to detect and identify targets in daylight and at night, using a combination of laser range finders, low-light high-resolution cameras, thermal cameras and infrared sensors. The system may be equipped with machine guns and an acoustic device that emits a tone powerful enough to make intruders nauseous and drop to the ground. The system can be operated manually, but also has an automatic mode in which it fires on its own.

30 Hellstrom Thomas, On the Moral Responsibility of Military Robots, available at: https://www8.cs.umu.se/~thomash/reports/ETIN2012.pdf, accessed11 October 2015.

31 Article 1 of the Universal Declaration of Human Rights (1948) states, "All human beings are born free and equal in dignity and rights." According to Kant, dignity means that the individual has an intrinsic worth, and dignity has no equivalent. Every human being, therefore, must be respected for his or her unique inherent or intrinsic value.

32 Johnson Aaron and Axinn Sidney, The Morality of Autonomous Robots, *Journal of Military Ethics*, Volume 12, Issue 2, August 2013.

33 Hellstrom Thomas, On the Moral Responsibility of Military Robots, available at: https://www8.cs.umu.se/~thomash/reports/ETIN2012.pdf, accessed11 October 2015.

34 Asaro Peter, On banning autonomous weapon systems: human rights, automation, and the dehumanization of lethal decision-making, *International Review of the Red Cross*, Vol. 94, No. 886, Summer 2012, p. 687–709.

35 The Manhattan Project, a secret military project, was launched in August 1942 to produce the first US nuclear weapon. In July 1945, the first test of a plutonium-based nuclear weapon was carried out in New Mexico. Just one year later, on 6 August 1945, 'Little Boy', a gun-type uranium-235 weapon, was dropped on Hiroshima. On 9 August 1945, 'Fat Man', an implosion-type plutonium-239 weapon, was dropped on Nagasaki. It has not been possible to make accurate estimates of the long-term fatalities in Hiroshima and

Nagasaki, given the large-scale destruction of records, population movements and the general censorship on nuclear effects by the US occupation regime. It is estimated that the effect of a 'limited' nuclear war involving 100 Hiroshima-sized bombs (i.e. less than half a per cent of the world's stockpile) would be disastrous for humanity. The five million tonnes of soot produced by the ensuing fires would cause the global temperature to fall by an average of 1.3°C. The disrupted global climate would have an overwhelming impact on food production. Nearly a billion people around the world could face starvation as a result of nuclear war.

36 The nuclear bomb presented a series of difficult legal questions. Generations of citizens, yet unborn, would be deeply affected by any nuclear detonation, how is one to apply the principle of proportionality and distinction? Hansen Emma, From nuclear bombs to killer robots: how amoral technologies become immoral weapons, Bulletin of the Atomic Scientists, 27 August 2014, available at: http://thebulletin.org/nuclear-bombs-killer-robots-how-amoral-technologies-become-immoral-weapons7409, accessed 13 July 2015.

37 The Norwegian Nobel Committee acclaimed US President Obama's vision and work for a world without nuclear weapons by awarding him the 2009 Peace Prize. Obama reaffirmed his vision and referred to it during an address in Berlin in 2013: "Peace with justice means pursuing the security of a world without nuclear weapons—no matter how distant that dream may be. And so, as President, I've strengthened our efforts to stop the spread of nuclear weapons, and reduced the number and role of America's nuclear weapons...."

38 Asaro Peter, On banning autonomous weapon systems: human rights, automation, and the dehumanization of lethal decision-making, *International Review of the Red Cross*, Vol. 94, No. 886, Summer 2012, p. 687–709.

VI LAWS: Legal Review

Introduction

A fully autonomous weapon (LAWS) will have the ability to "search for, identify and attack targets, including human beings, using lethal force without any human operator intervening."[1] Though LAWS in the true sense are not yet available in the military arsenal, a number of countries, including China, France, Germany, India, Israel, South Korea, Russia, the UK and the United States are developing such capabilities.[2] Experts predict that fully autonomous lethal weapon systems could be developed in the next 20 years. Then LAWS could mean a revolutionary change for the States which have them, bringing down casualties of combatants and the cost of war, but for those against which they would be deployed, they would mean greater casualties among non-combatants and civilians, including women and children. A case in point is the use of lethal drones by the US military in Afghanistan and the tribal areas of Pakistan.

A large section of robotic engineers, ethical analysts, and legal experts are of the firm belief that LAWS will never be able to select and strike targets on the basis of an analysis of a complex situation in an armed conflict. They will never meet the standards of distinction and proportionality required by international law, nor be capable of understanding human nuances and being true to the tenets of mercy, identification, and morality. How would an autonomous weapon distinguish between a military doctor who is armed for self-defence and a combatant, for example? Another problem with the use of LAWS would be in fixing responsibility in case of violations of IHL.[3] Besides, once LAWS become available, they may be used by terrorist organizations and rogue armies.

Another section of people believe that it is pointless to debate on LAWS at this juncture as the full potential of such a system is yet to be realized.[4] These people feel it is possible that in the future, LAWS will be able to identify targets better than human beings and respond more rapidly and accurately and cause less collateral damage. However, postponing the debate may render it meaningless as it will be too late to impose effective restrictions on the development of LAWS. In addition, a government which would have spent a huge amount on research and development would hesitate to sign any treaty prohibiting or restricting the use of LAWS.

Legal Review of Weapons

Over the last 150 years or so humanity has, on the one hand, used technological advancement to develop more destructive weapons, and on the other, made greater efforts to limit or regulate weapons that may cause aggravated suffering to combatants. The 1868 Declaration of St Petersburg was the first formal agreement where the obligation to review new technologies was recognized at an international forum. Its preamble established the principle that the only legitimate object of war is to weaken the military forces of the enemy and that this purpose would be exceeded by "the employment of arms which uselessly aggravate the sufferings of disabled men, or render their death inevitable". The Declaration laid emphasis on reviewing of the legality of new weapons. It stated: "The Contracting or Acceding Parties reserve to themselves to come hereafter to an understanding whenever a precise proposition shall be drawn up in view of future improvements which science may effect in the armament of troops, in order to maintain the principles which they have established, and to conciliate the necessities of war with the laws of humanity".[5]

The Declaration affirmed that the employment of "explosive projectiles of less than 400 gm in weight which is either explosive or charged with fulminating or inflammable substances," would be contrary to the laws of humanity and specifically prohibited the use of such weapons in armed conflict.[6] It prompted the adoption of further declarations of a similar nature at the two Hague Peace Conferences of 1899 and 1907. The 1899 Hague Declaration (IV, 2) prohibited "the use of projectiles whose sole object is to diffuse asphyxiating or deleterious gases". The prohibition was derived from the general principles of customary international law prohibiting the use of poison and materials causing unnecessary suffering.[7] During the

First World War, the use of gases began with irritant gases, but escalated rapidly. After the War, the prohibition of gas warfare was reaffirmed in the 1919 Treaty of Versailles and the 1925 Geneva Protocol.[8] The 1899 Hague Declaration (IV, 3) prohibited the use of Dum-Dum bullets, which expanded or flattened easily in the human body making wounds uselessly cruel.[9] The Declaration has been regarded as codifying one aspect of the customary rule prohibiting weapons causing unnecessary suffering.[10] The Rome Statute of the International Criminal Court includes in its list of war crimes the use of expanding bullets.[11] The 1907 Hague Convention VIII regulated the laying of automatic submarine contact mines. The treaty prohibited: (i) laying of unanchored automatic contact mines, except when they are so constructed as to become harmless one hour at most after the person who laid them ceases to control them; (ii) laying of anchored automatic contact mines which do not become harmless as soon as they have broken loose from their moorings; and (iii) using torpedoes which do not become harmless when they have missed their mark. The Convention also prohibited the laying of automatic contact mines off the coast and ports of the enemy, with the sole object of intercepting commercial shipping.[12] At the close of the war, the Contracting Powers were to remove the mines which they had laid.

Such efforts at regulating or banning harmful weapons and methods of warfare have been continuing and the list includes biological and chemical weapons, environmental modification techniques, anti-personnel landmines, and cluster munitions. In addition, the 1980 UN Convention and its four Protocols prohibit or restrict the use of conventional weapons (i) whose primary effect is to injure by fragments not detectable in the human body by X-rays, (ii) landmines, booby-traps and other devices, (iii) incendiary weapons, and (iv) laser weapons whose combat function causes permanent blindness. The Fifth Protocol of the 1980 Convention relates to the clearance, removal or destruction of explosive remnants of war. Today, parties to an armed conflict are limited in their choice of weapons and methods of warfare by the rules of IHL governing the conduct of hostilities. The binding customary rules include the prohibition of means and methods of warfare of a nature that causes superfluous injury or unnecessary suffering and of means of warfare that are incapable of distinguishing between civilians or civilian objects and military targets.

Review of LAWS as New Weapons

Besides the provision contained in The St Petersburg Declaration, Article 36 of AP I refers to the need to review of new weapon technologies.[13] In the context of IHL, an autonomous weapon system can be classified as a means of warfare and is therefore governed by Article 36. Before starting a discussion on the need for the review of LAWS as a 'new weapon', it is important to note that the right of the parties to any armed conflict to choose methods or means of warfare is not unlimited.[14] This notion of limited rights along with the coordinate principle of humanity is the bedrock of IHL. Article 35(2) of AP I prohibits the use of weapons, projectiles and material and methods of warfare of a nature that causes superfluous injury or unnecessary suffering. In order to limit collateral damage and calamities during an armed conflict, international law dictates that it is impermissible to use certain weapons regardless of the enemy or the context. A State's compliance with the international norms is determined by a mandatory review that considers two important qualities: (i) whether the weapon is capable of distinguishing legitimate military targets from civil objects and persons, and (ii) whether it causes superfluous injury or unnecessary suffering.

Article 36 codifies the requirement to conduct legal reviews of all new weapons. The obligation to conduct reviews applies to every state party to AP I, whether it develops, modifies, manufactures or acquires a new weapon system. Even States which are not a party to AP I have such an obligation under the rules of customary international law. The International Court of Justice, in its Advisory Opinion on nuclear weapons has identified certain cardinal principles of humanitarian law as being customary. These include the principle of distinction and the ban on employing weapons that are incapable of distinguishing between civilian and military targets; and the prohibition of causing unnecessary suffering to combatants.[15] It is worth mentioning that a few States that are not yet party to AP I have adopted national procedures to ensure that their weapons are subject to this type of review.

Article 36 provides:

> In the study, development, acquisition or adoption of a new weapon, means or method of warfare, a High Contracting Party is under an obligation to determine whether its employment would, in some or

all circumstances, be prohibited by this Protocol or by any other rule of international law applicable to the High Contracting Party.

Article 36 is contained in part III of the 1977 AP I dealing with the law of war, and particularly that part of the law of war which relates to methods and means of warfare. It is linked to Article 35 (Basic rules) of AP I, and the introduction of new weapons by the States. Article 36 implies the obligation to establish internal procedures for the purpose of elucidating the issue of legality.[16]

A question which arises in the case of the use of LAWS is whether a State has unlimited rights to use any weapon in an armed conflict. Article 35 (1) states that "In any armed conflict, the right of the parties to the conflict to choose methods or means of warfare is not unlimited." The use of the term "in any armed conflict" means this is applicable to all conflicts, whether international or non-international. Therefore, the parties to a conflict are not free to use any method or means of warfare whatsoever.[17] The 1907 Hague Regulations states in Article 22, "The right of belligerents (parties to an armed conflict) to adopt means of injuring the enemy is not unlimited." In 1965 the International Conference of the Red Cross reaffirmed this principle stating, "The right of Parties to the conflict and of members of their armed forces to adopt methods and means of combat is not unlimited." In international law there are no exceptions to this fundamental rule.

In addition to the general principle by which the right of belligerents to adopt means of injuring the enemy is not unlimited, there are two other fundamental rules: humanitarian rules and the rules on good faith. The humanitarian rules prohibit the killing or wounding of an enemy who has laid down his arms or no longer has the means to defend himself and has therefore surrendered unconditionally. They also prohibit refusing to give quarter and causing superfluous injury or unnecessary suffering. The rules on good faith prohibit killing or wounding the enemy treacherously. The ICRC commentary states that the armed forces must include these fundamental principles in the instructions they issues to their troops.[18] It is feared that LAWS would not be in a position to follow these two fundamental rules.

The ICRC's Guide to the "Legal Review of New Weapons, Means and Methods of Warfare" explains that Article 36 aims to prevent the use of weapons that would violate international law in all circumstances and to impose restrictions on the use of weapons that would violate international law in some circumstances, before they are developed or acquired by the armed forces of a State.[19] Article 36 is complemented by Article 82 of AP I, which requires that legal advisers be available at all times to advise military commanders on IHL and "on the appropriate instruction to be given to the armed forces on this subject". Both provisions establish a framework for ensuring that the armed forces will be capable of conducting hostilities in strict accordance with IHL through legal reviews of planned means and methods of warfare.

The Scope of Review under Article 36

The scope of Article 36 is very broad. It covers weapons of all types, anti-personnel or anti-materiel, lethal, nonlethal, and weapons systems. Therefore, all autonomous systems likely to be used by the armed forces, including lethal ones and those which could be modified into lethal versions would come under the purview of Article 36, as would the ways in which these weapons are to be used pursuant to military doctrine, tactics, rules of engagement, operating procedures and counter measures. The weapons for review would include all weapons purchased, to be acquired, modified, or developed as per military specifications. An existing weapon that is modified in a way that alters its function, or a weapon that has already passed a legal review but that is subsequently modified is also needed to be reviewed. If a state has joined a new international weapon treaty, it may require to recheck the legality of the existing weapon. In case, there is any doubt as to whether the device or system proposed for study, development or acquisition is a 'weapon', legal advice must be taken from the authority responsible for review of the weapons.

The obligation to review the legality of new weapons implies at least two things: (i) A State should have in place some form of permanent standing mechanism that can be activated at any time when it is developing or acquiring a new weapon; (ii) Such a procedure should be made mandatory for the authority responsible for developing or acquiring new weapons. Other than these minimum procedural requirements, a State is free to decide the other specific necessities of its review mechanism.[20]

The legality of a weapon does not depend solely on its design or intended purpose, but also on the manner in which it is expected to be used in an armed conflict. A weapon, therefore, cannot be assessed in isolation from the method of warfare by which it is to be used. For instance, a weapon when used in a particular manner could pass the test and be certified 'lawful' under Article 36, but it may not be so when used in a different manner. This is why Article 36 requires a State to determine whether the employment of weapon would, in some or all circumstances, be prohibited by international law. Article 8 of the UN Convention on Certain Conventional Weapons (CCW) provides for a review mechanism with the purpose of examining new categories of conventional weapons which do not yet fall under the Protocols annexed to the Convention. A Conference will be held when this is requested by a Contracting Party to the Convention with the agreement of a majority of at least 18 of them. All States, including all Parties to the Protocol, will be invited to this Conference. Thus, even if an autonomous weapon system is inducted into service after a review under Article 36, further testing and review may be conducted to gain an additional insight into its capabilities and to ensure that the weapon system actually meets the requirement.[21]

Weapon Review and State Practices

The importance of the obligation to review a weapon under Article 36 cannot be overstated. The States are bound to undertake a legal review of new weapons and technologies of warfare. Failure to do so would render a State internationally responsible for a breach of its obligations under AP I.[22] The review mechanism followed in Australia, Germany, Norway, Sweden, the UK and the US will be discussed in the following sections. Incidentally, the US is not a party to AP I, but has a mechanism to undertake the legal review of new weapons.

Australia

The Director General Australian Defence Force Legal Services (DGADFLS) is mandated to review all proposed new weapons to determine whether the weapons or their intended use in combat are consistent with the obligations assumed by the Australian Government under applicable treaties to which Australia is a party and customary international law. [23] In cases where the Director of Military Operations (DMO) is responsible for the study, development, acquisition, adaptation or modification of a weapon outside the capability development process, the DMO is to seek a legal review of the weapon from the DGADFLS.

The Defence Instructions (General) requires a legal review of all proposed new weapon acquisitions to determine whether their intended use is consistent with the Australian Government's obligations under international law. These instructions state that the Australian Defence Force (ADF) members must take care to ensure that the weapons are used and employed in a manner that complies with the law of armed conflict (LOAC). It clarifies that the use of a weapon will be unlawful under the LOAC if it breaches the principle of proportionality by causing unnecessary injury or suffering. In a major or extended conflict, ADF members could be called upon to utilize captured enemy weapons. While the LOAC recognizes that such weapons may be used (after enemy markings are removed and provided they do not cause unnecessary injury or suffering), prior command approval should normally be obtained if the captured weapon is not currently in the ADF inventory. [24]

In order to facilitate reviews under Article 36, detailed information on the weapon under review is required. Such information is generally obtained from the manufacturer and other armed forces or through specialized literature, expert opinions or other credible sources. In order to assess the legality of new weapons, a list of questions is also provided by the DGADFLS: (i) What is the purpose of the new weapon? (ii) What are the factors which favour the introduction of the new weapon? (iii) What is the damage mechanism of the new weapon—blast, fragmentation, etc.? (iv) Is the new weapon specifically designed to cause injury to personnel? (v) What human injuries will the new weapon be capable of inflicting? (vi) What other weapons, if any, would be capable of fulfilling the same purpose as the new weapon? (vii) Has the new weapon been adopted by the armed forces of other States or by other agencies in Australia or overseas and if so, by which ones? (viii) Is evaluation data concerning the new weapon available from the armed forces of other States or from other agencies in Australia or overseas? In addition to these questions, the DGADFLS is provided with guidelines indicating how to assess the information gathered and outlining the appropriate legal criteria. Examples of past assessments are also included. Although the final decision is not legally binding, a negative outcome would prevent the acquisition of the weapon, provided there is no contradiction in the legal opinion. Although there is no formal appeal process, further advice may be sought before proceeding with the acquisition. The DGADFLS maintains a comprehensive record of all weapons reviews.[25]

Germany

In order to ensure the implementation of its obligation under Article 36 of AP I, Germany has established a Steering Group on Review of New Weapons and Methods of Warfare within the Federal Ministry of Defence. The Steering Group is under the Directorate-General for Legal Affairs' International and Operational Law Branch. Representatives of all other competent Directorates-General of the Ministry of Defence are part of the Steering Group to synergize the in-house knowledge of all experts, ranging from political to technical or operational expertise. The list includes the Directorates-General for Security and Defence Policy; Equipment, Information Technology and In-Service Support; Planning; Forces Policy; and Strategy and Operations. The representatives of the Directorates-General are points of contact through whom the Directorate-General for Legal Affairs can obtain subject matter expertise for a weapon review. They may also bring in projects for review on behalf of their Directorate-General. The Steering Group is a permanent structure. The representatives of the competent Directorates-General may vary depending on the matter under review.

The legal criterion for the review is primarily the IHL as applicable to Germany. Article 36-reviews are initiated at the earliest possible stage of a new-weapon-project. The Steering Group assesses whether the employment of the weapon, means or method of warfare under review would, in some or all circumstances, be prohibited by AP I or by any other rule of international law applicable to Germany. The Steering Group's assessment is a recommendation within the development and procurement process, and not a final decision about the introduction of a weapon, means or method of warfare. Thus, it has neither a binding character nor can it be appealed through a specific procedure. The review results in a finding and the recommendations are recorded. Questions of accessibility are decided on a case-by-case basis pursuant to applicable domestic law. Depending upon the complexity of the subject, the review process might be phased in accordance with respective development steps.

The Steering Group, besides receiving information from the industry, may also seek additional information from outside or within the armed forces. Tests and evaluations are conducted throughout the procurement process, which is supervised by an Integrated Project Team (IPT) that is created and maintained for the entire life cycle of the product. The IPT's role is to ensure the continuous availability of know-how, and uninterrupted and smooth cooperation of all parties involved in the procurement and in-service process at the Ministry of Defence. Once a new product has been validated for procurement, the Directorate-General for Equipment, Information Technology and In-Service Support appoints a project manager as head of the IPT. The project manager represents the project externally, for example by producing reports for the committees of the German Bundestag, the Federal Audit Office, or other institutions or steering boards, and is responsible for integrated compliance demonstrations, including operational testing.[26]

The German Government is of the opinion that the regulations introduced by AP I apply only to conventional weapons; and dual-use systems fall outside the scope of application of Article 36. However, they may be considered for review if it can be established that their intended use clearly contributes to the conduct of warfare. The assessment of whether a system or device should be subject to a review is made on a case-by-case basis. Germany is of the view that given the actual state of the art of artificial intelligence and other important components of LAWS, a legal review would for the time being lead to LAWS being illegal, as they would not be able to meet the requirement set out by Article 36.[27]

Norway

Norway ratified AP I on 27 November 1981. In 1994, the Department of Defence issued a directive for the implementation of the obligations under Article 36. The term 'weapon' in this directive means any means of warfare, weapon system/project, substance etc. which is particularly suited for use in combat, including ammunition and similar functional parts of a weapon.[28] A permanent advisory committee, called the Chief of Defence International Law Committee (FFU), was also established for the evaluation of the legal aspects of new weapons, means and methods of warfare.[29] The FFU on its own initiative or on the basis of inquiries from units within the Defence Military Organization, undertakes assessments. In addition, the Chief of Defence Staff (CDS) may request the committee to assess any other matter within the committee's area of competence.

The CDS provides advice and prepares reports on important issues related to Norway's obligations under Article 36 and to the legal review of weapons, means and methods of warfare in international operations taking place in situations other than armed conflict. It ensures that relevant units within the Defence Military Organization assess international legal aspects in connection with studies, development, acquisition or approval of new weapons, methods or means of warfare. This includes the legal review of existing weapons, methods and means of warfare, in particular when Norway commits to new international legal obligations.

Legal review is supposed to be made as early as possible, in the concept/study phase, when the operational needs are identified, the military objectives are defined and the technical resources and financial conditions are settled. If the circumstances at a later stage change significantly, the international legal aspects are to be reassessed. The methods and means of warfare are normally established through guidelines on operative planning and through rules on the use of force (rules of engagement). Assessment of international law is incorporated in the planning process, the descriptions of operative planning and in the manuals for operational assessment. In addition, systems supporting the operative planning process are assessed according to international law. When acquisitions, development projects, etc. are forwarded to the Ministry of defence for approval, the case documents clarify that the matter has been assessed according to international law, unless it falls outside the scope of Article 36. The review is based on the existing international law (treaty law and international customary law) by which Norway is bound. The Norwegian Defence Research Institute reports to the Ministry of Defence with regard to participation in international research and development programmes that may have international legal obligations.[30] The FFU is required to submit an annual written report to the CDS on the activities of the committee and inform the International Humanitarian Law Contact Committee on its work.

Sweden

Sweden was the first country to establish an independent decision-making body for reviewing the legality of weapons. As early as in 1974, Sweden established the Delegation for International Law Monitoring of Arms

Projects. The decision to establish the delegation was made after Sweden's critical assessment of the use of certain weapons during the Vietnam War by parties to that conflict, and stemmed from its desire to ensure that the means and methods of warfare used by the Swedish Armed Forces were in compliance with its international obligations.[31]

The weapon review mechanism is currently regulated through the Swedish Ordinance on International Law Review of Arms Projects, which requires the Swedish Armed Forces, the Swedish Defence Material Administration, the Swedish Defence Research Agency and other agencies to report all weapon projects to the delegation. The members of the delegation are elected by the Swedish Government. The delegation is an independent body with a status equivalent to a government authority and is not part of the government. The delegation conducts around two or three weapon reviews annually. It has to present an annual report on its activities to the government. It monitors planned purchases or modifications of all types of weapons (including non-lethal weapons) by the Swedish Armed Forces, Coast Guard and the Police Authority. It also reviews new military means and methods of warfare. It may review weapons bought by the Armed Forces without the involvement of the Material Command of the Armed Forces.

The delegation monitors planned purchases or modifications of weapons and issues approval or non-approval decisions. If a weapon project assessed by the delegation does not meet international law requirements (i.e., IHL, international human rights law and disarmament law), the delegation encourages the authority that submitted the weapon for review to take appropriate measures to bring the weapon in line with the requirements of international law. The decisions of the delegation are, however, not legally binding. They are only advisory to the authority that submitted the matter for review. The authority that requested the review can appeal the decision of the delegation to the Swedish Government. Under the Swedish principle of public access to official documents, it is possible to request access to the record of decisions and to official documents that are not classified.

The United Kingdom

In the UK, legal review is undertaken routinely in respect of weapon systems brought on to the UK inventory following the State's ratification of AP I on 28 January 1998. The weapons review process is conducted by the Ministry of Defence (MoD) in a progressive manner as concepts for new means and methods of warfare are developed and as the conceptual process moves towards procurement. The review process considers not only the law as it stands at the time of the review but also attempts to take into account likely future developments in the law of armed conflict.[32]

The review of all new weapons, means and methods of warfare is conducted by the Development, Concepts and Doctrine Centre (DCDC), through a team of qualified military lawyers from the air force, army and navy. The term 'weapon' is defined in its broadest sense. Weapons modified for a different use are also reviewed. Weapons are reviewed with regard to their design and intended use. [33] Provisions of IHL are central to the assessment of all weapon systems. Other international conventions and human rights law to which the UK is party are also considered.

The time frame of the review is dependent on the context. Reviews can be fast-tracked when an expedited decision is needed. However, a weapon review can last as long as the actual weapon development cycle. The reviewers consider all pertinent documentation provided by the manufacturer and the armed forces. The MOD may conduct additional independent tests and evaluations to verify the information supplied by the manufacturer. Reviews are conducted in consultation with equipment project and procurement teams, medical experts, government scientists, armed forces experts, environmental specialists and commercial and engineering companies that design and build the equipment. Each process is tailored to the specific weapon and the review requirements of that weapon.[34]

A legal review generally leads to a written legal advice. The military lawyers sign the review, but the process has joint ownership. The officials involved in the review process, in particular, the experts consulted during the process, have to confirm that all the information reported in the written legal advice is correct before it is signed. The legal advice is then peer-reviewed by another lawyer within the DCDC. Nuclear weapons are an exception to the review process.[35]

The United States

In 2015, the US Department of Defence (DoD) issued a thoroughly revised Law of War Manual. The Manual states that the acquisition and procurement of weapons and weapon systems shall be consistent with all applicable domestic laws, treaties and international agreements, IHL and customary international law. An attorney authorized to conduct such legal reviews shall conduct the legal review of the intended acquisition of weapons or weapons systems.[36] The legal review may approve (with or without conditions) or disapprove of the weapon. In the latter case, the review can indicate that a favourable legal review may be possible on reconsideration, provided the corrective actions identified in the review are taken. All reviews are documented but the records are classified. When the system is exported to another country, some information from the review can be provided.

The Manual makes a clear distinction between means and methods of warfare. 'Means of warfare' refers to the intended effect of weapons in their normal and expected use against combatants, while 'methods of warfare' refers to the employment of weapons in a broader sense. The review of the acquisition or procurement of a weapon should consider three questions: (i) whether the weapon's intended use is calculated to cause superfluous injury; (ii) whether the weapon is inherently indiscriminate; and (iii) whether the weapon falls within a class of weapons that has been specifically prohibited.[37] If the weapon is not prohibited, the review should also consider whether there are legal restrictions on its use that are specific to that type of weapon. If any specific restrictions apply, then the intended concept of employment of the weapon should be reviewed for consistency with those restrictions.[38] Any captured or other foreign weapons are not be used in combat unless they have undergone the required legal review and have been duly issued to personnel. The Law of War Manual, however, does not specifically mention any policy guidelines for the review of LAWS likely to be developed by the US in the future.

The legal review process is an ongoing exercise throughout requirement identification, weapon system development, testing and evaluation, fielding, and employment. There is significant coordination between the reviewing authority, the weapon providers, the procurement agency and the end-user. Others who may also participate in reviews include safety officers, software developers, lawyers and engineers. An initial review of each weapon or weapon system is made after the formal request for review is placed and before the stage of formal development. A final legal review must be made prior to the award of the initial contract for production. The review may take into account feedback from users once the weapon has been deployed.

The DoD Directive (Number 3000.09) of 2012 lays out guidelines for the development and use of autonomous and semi-autonomous weapon systems. The Directive represents the first policy announcement by any country on fully autonomous weapons, which do not yet exist but would be designed to select and engage targets without human intervention. It highlights the weaknesses of fully autonomous weapons. Its glossary contains an extensive list of the possible causes of failures in these and other autonomous weapons: human error, human-machine interaction failures, malfunctions, communications degradation, software coding errors, enemy cyber attacks or infiltration into the industrial supply chain, jamming, spoofing, decoys, other enemy countermeasures or actions, or unanticipated situations on the battlefield.[39] The Directive states that before a decision is taken to enter into the formal development of a semi-autonomous system, a preliminary legal review must be completed in coordination with the General Counsel of the DoD… and in accordance with the relevant policy guidance.[40]

The Directive (Number 3000.09) acknowledges that fully autonomous weapons could endanger civilians in many ways.[41] It recognizes that failures may occur and lead to unintended engagements.[42] Paragraph 4(e) of the Directive allows for international sales and transfers. It states that transfers must be approved "in accordance with existing technology security and foreign disclosure requirements and processes", but once these weapons leave the country, the US would lose exclusive control over them.

The Challenges of Review under Article 36

The requirement that the legality of all new weapons, means and methods of warfare be systematically assessed is arguably one that applies to all States, regardless of whether they are party to AP I.[43] The obligation to conduct legal reviews of new 'means of warfare' before their use is generally considered a reflection of customary international law. However, consensus is lacking on whether an analogous requirement exists in the case of 'methods of warfare'. According to Schmitt (2013), a state party to AP I is required to conduct a legal review of autonomous weapon systems and the methods of their use in an armed conflict. The States which have not ratified AP I also have customary obligation to undertake such reviews. Both reviews, whether or not legally mandated, are well-advised whenever feasible.[44] The ICRC too has emphasized the importance of the legal review of weapons in a number of international conferences.[45] Carrying out a legal review of autonomous weapons is of particular importance today in order to prevent the use of such weapons in an illegal manner.

There are a few shortcomings in Article 36. First, it does not specify how the legality of weapons, and means and methods of warfare is to be determined. A number of States, while ratifying AP I, have added reservations to exempt certain types of weapons from the scope of Article 36.[46] For example, Germany, applies the provisions exclusively to conventional weapons.[47] Also, equipment of a dual-use nature (for example, an autonomous system which can be used in a non-lethal as well as lethal mode) would not be subject to review unless it can be determined that it directly contributes to the conduct of warfare. Another problem is that reviews undertaken by States remain national procedures beyond any kind of international oversight, and there are no established standards on how they should be conducted. According to Thomas Nash, the review process of weapons is inadequate. Unlike the rigorous testing process used to develop drugs, cars and most consumer goods, there is virtually no public scrutiny of the process of developing and using new weapons.[48] It is most likely that the development of LAWS will remain a closely guarded secret as States seek to establish a military advantage over their perceived adversaries.[49]

The system of legal review of new weapons must be trustworthy and devoid of any ambiguity. According to Sharkey (2008), in the case of the MQ1-Predator UCAV, the Judge Advocate General's Corps (JAG) first passed it for surveillance missions. When the Predator was armed with Hellfire missiles later, JAG held that the combination did not require a review because both the Predator and Hellfire missiles had been reviewed earlier..[50] A scrupulous weapon review is therefore crucial to ensuring that autonomous weapon systems are developed, produced, fielded and used in compliance with international law. The legal review of autonomous military systems must also be guided by the principles of humanity and the dictates of public conscience, as set out in the Martens Clause in the preamble to the 1899 Hague Convention II, the preamble to the 1907 Hague Convention IV, and Article 1(2) of AP I.[51]

The International Court of Justice in its 1996 Advisory Opinion has elaborated the criteria for evaluating the legality of new weapons:[52]

> The cardinal principles contained in the texts constituting the fabric of humanitarian law are the following. The first is aimed at the protection of the civilian population and civilian objects and establishes the distinction between combatants and non-combatants; States must never make civilians the object of attack and must consequently never use weapons that are incapable of distinguishing between civilian and military targets. According to the second principle, it is prohibited to cause unnecessary suffering to combatants: it is accordingly prohibited to use weapons causing them such harm or uselessly aggravating their suffering. In application of the second principle, States do not have unlimited freedom of choice of means in the weapons they use.

Technical Challenges

New technology is likely to make the weapon review process more complex,[53] as it may pose specific risks, and require new and specific methods of assessment. The increasing sophistication of weapon systems and the integration of hardware and software, would also increase the cost of the review process. Many weapon-developing States may not even be equipped to deal with the review process due to the lack of experience, technical expertise, or financial resources.

At present, there is no common understanding on the definition of LAWS. The States developing such systems and civil society organizations have different understandings of the capabilities required for a system to be considered autonomous, and accordingly they may not agree on the classification of existing systems. Reviewing the legality of weapons with automated and autonomous features also presents a number of technical challenges as such weapons demand complex procedures for testing weapon performance and evaluating the risks associated with unintended loss of control. As such assessments require significant technical and financial resources, there is a strong incentive for deepening cooperation and information sharing between States in the area of weapon reviews.

Increased interaction can facilitate the identification of best practices and ways of reducing costs associated with testing and evaluation procedures.[54]

In order to comply with IHL, LAWS would have to be capable of following the rules of distinction and proportionality. The key issues which could be considered in a legal review of autonomous military systems are as follows: A. Technical characteristics, capabilities and intended effects: (i) Is the system capable of compliance with international law? (ii) Could the use of the system in normal conditions cause unnecessary suffering or superfluous injury to combatants? (iii) Could the system cause long-term, widespread or severe damage to the natural environment? (iv) In case the system is lethal, can it select and fire at a target autonomously and can it comply with the principles of distinction, proportionality and precaution in attack? (v) Has automation improved distinction, proportionality and precaution in the application of force in contrast to existing weapons? B. Lethality and Human Rights issues: (i) Does use of the weapon constitute a violation of the right to life or the right to dignity of the target? (ii) Could it be considered unacceptable under the principles of humanity and the dictates of public conscience? (iii) Could it create a responsibility gap in the case of a violation of international law? Depending on the response to these questions, the weapon review committee constituted by a State could place restrictions or make recommendations on the use of the system, which could be integrated into the rules of engagement. [55]

Conclusion

The States Parties to AP I are obliged to determine the legality of autonomous weapon systems before adapting, acquiring, or developing them. All States are bound by customary international law to ensure that new weapons acquired by them comply with those rules.[56] They are prevented from incorporating such weapons into their arsenal as would violate international law if used.[57] However, the weapon review process is limited to a few states. The ICRC's attempt to assist states to establish an evaluation process to meet the obligations in Article 36 has not been successful.[58] Besides, the treaty itself (AP I) does not have universal ratification.[59] Attempts by various international NGOs and individual states to propose an independent "international scrutinizing mechanism" have never met with support.[60] It is doubtful that the development and procurement of autonomous weapons systems by states and their intended use during armed conflicts would be consistent with their treaty obligations. The traditional interpretation of treaty law rules may not apply to autonomous weapon systems and future weapons made possible by technological advancement.[61]

Endnotes

1   Autonomous Weapons: States Must Address Major Humanitarian, Ethical Challenges, Geneva: International Committee of the Red Cross, September 2013, available at: http://www.icrc.org/eng/resources/documents/faq/q-and-a-autonomous-weapons.htm., accessed 12 September 2015.

2   Besides the US, there are 43 nations that are building, buying and using military robotics today. Jensen Eric Talbot, The Future of the Law of Armed Conflict: Ostriches, Butterflies, and Nanobots, *Michigan Journal of International Law*, Vol. 35, No. 2, 2014, p. 288.

3   When an autonomous weapon system kills an innocent person unlawfully, whom shall we try: the weapon system, its designer, its programmer, the soldier who deployed it, or the military commander who decided that it could be used in the armed conflict? A U S Department of Defence directive provides: "Persons who authorize the use of, direct the use of, or operate autonomous and semi-autonomous weapon systems must do so with appropriate care and in accordance with the law of war, applicable treaties, weapon system safety rules, and applicable rules of engagement (ROE)." The US Defence (DoD) Directive 3009.09, 21 November 2012, paragraph 4b.

4   Vallejo Daniel, Electric Currents: Programming Legal Status into Autonomous Unmanned Maritime Vehicles, *Case W. Res. J. Int'l L.* , Vol. 47, (2015), p. 405-428.

5   The 1868 St Petersburg Declaration.

6   The St Petersburg declaration had its origin in the invention, in 1863, by Russian military authorities of a bullet which exploded on contact with a hard substance and whose primary object was to blow up ammunition wagons. In 1867 the projectile was so modified as to explode on contact with a soft substance. The Russian Government, unwilling to use the bullet itself or to allow another country to take advantage of it, suggested that the use of the bullet be prohibited by international agreement. The Declaration to that effect adopted in 1868, which has the force of law, confirms the customary rule according to which the use of arms, projectiles and material of a nature to

cause unnecessary suffering is prohibited.

7   These general customary principles were embodied in Articles 23 (a) and 23 (e) of the Regulations annexed to 1899 Hague Convention II and 1907 Hague Convention IV.

8   The 1925 Geneva Protocol for the Prohibition of the Use in War of Asphyxiating, Poisonous or Other Gases, and of Bacteriological Methods of Warfare.

9   The British army had modified a rifle bullet, manufactured in Dum-Dum in India. These were claimed to be highly effective in their wars against "active and brave barbarian foes". The Sub-Commission to the First Commission to the 1899 Hague Peace Conference perceived Dum-Dum bullets as having similar effects to a projectile that carried explosive material. These bullets caused a very small hole at the entry point, while the exit wound was very large, causing enormous ravages in the body. The British delegate at the conference argued that there was a difference in war between civilized nations and that against savages. The Sub-Commission subsequently proposed that "The use of bullets which expand or flatten easily in the human body (making wounds uselessly cruel), such as explosive bullets, bullets with a hard envelope which does not entirely cover the core or is pierced with incisions, ought to be prohibited." Coupland Robin and Loye Dominique, The 1899 Hague Declaration concerning Expanding Bullets: A treaty effective for more than 100 years faces complex contemporary issues, *International Review of the Red Cross*, Vol. 85, No. 849, March 2003, p. 135-142.

10   Roberts Adam and Guelff. 2000. *Documents on the Laws of War*, Oxford: Oxford University Press, p. 63.

11   The Statute of the International Criminal Court, 1988, Article 8 (2)(b)(xix).

12   1907 Hague Convention VIII Relative to the Laying of Automatic Submarines Contact Mines, Articles 1 and 2.

13   Unfortunately, the provision for review of new weapon technologies contained in Article 36 has not been given the required attention or importance by most States. Gaoust Isabelle, Robin Coupland and Rikkie Ishoey, New Wars, New Weapons: The Obligations of the States to Assess the Legality of means and methods of Warfare, *International Review of the Red Cross*, Vol. 84, No. 846, June 2002, p. 345-362.

14   Article 35 (1), the 1977 Additional Protocol I to the Geneva Conventions of 12 August 1949 and Relating to the Protection of Victims of International Armed Conflicts.

15   The International Court of Justice, *Legality of the Threat or Use of Nuclear*

Weapons, Advisory Opinion, 1996 ICJ 226 (8 July 1996). The Court also made the important declaration that the States do not have unlimited freedom of choice of means in the weapons they use.

16 Sandoz Yves, Chrisotphe Swinarski and Bruno Zimmermann (eds.). 1987. *Commentary on the Additional Protocols of 8 June 1977 to the Geneva Conventions of 12 August 1949*, International Committee of the Red Cross, Geneva: Martinus Nijhoff Publishers, p. 424.

17 Sandoz Yves, Chrisotphe Swinarski and Bruno Zimmermann (eds.). 1987. *Commentary on the Additional Protocols of 8 June 1977 to the Geneva Conventions of 12 August 1949*, International Committee of the Red Cross, Geneva: Martinus Nijhoff Publishers, p. 390.

18 The rules on good faith prohibit killing or wounding the enemy .treacherously, as well as deceiving him by the improper use of the flag of truce, of national emblems or of enemy uniforms, and also by the improper use of the Red Cross emblem. Every military Power, without exception, must include these fundamental principles in the instructions it issues to its troops. Sandoz Yves, Chrisotphe Swinarski and Bruno Zimmermann (eds.). 1987. *Commentary on the Additional Protocols of 8 June 1977 to the Geneva Conventions of 12 August 1949*, International Committee of the Red Cross, Geneva: Martinus Nijhoff Publishers, p. 382.

19 Lawand Kathleen *et al.*, *Guide to the Legal Review of New Weapons, Means and Methods of Warfare: Measures to Implement Article 36 of Additional Protocol I*, Geneva: ICRC, 2006, p. 4.

20 Lawand Kathleen, Reviewing the legality of new weapons, means and methods of warfare, Vol. 88, No. 864, *International Review of the Red Cross*, December 2006, p. 927.

21 Backstrom Alan and Henderson Ian, New Capabilities in Warfare: An Overview of contemporary technological developments and associated legal and engineering issues in Article 36 weapons review, *International Review of the Red Cross*, Vol. 94, No. 886, Summer 2012, p. 510.

22 Rappert Brian, Richard Moyes, Anna Crowe and Thomas Nash, The roles of civil society in the development of standards around new weapons and other technologies of warfare, *International Review of the Red Cross*, Vol. 94, No. 886, Summer 2012, p. 779.

23 The Australian Defence Instructions (General) OPS 44–1—Legal Review of New Weapons.

24 The Law of Armed Conflict, Executive Series, ADDP 06.4, 11 May 2006, paragraph 4.2 and 4.3.

25 Daoust Isabelle, Robin Coupland and Rikke Ishoey, New Wars, New Weapons: The Obligation of States to Assess the Legality of Means and Methods of Warfare, *International Review of the Red Cross*, Vol. 84, No. 846, June 2002, p. 345-362.

26 Boulanin Vincent, Implementing Article 36 Weapon Reviews in the Light of Increasing Autonomy in Weapon Systems, SIPRI Insight on Peace and Security, No. 2015/1, November 2015.

27 Statement on transparency made by the Federal Republic of Germany at CCW Expert Meeting on LAWS at Geneva, 13-17 April 2014, available at: https://www.unog.ch/80256EDD006B8954/(httpAssets)/07006B8A11B9E93 2C1257E2D002B6D00/$file/2015_LAWS_MX_Germany_WA.pdf, accessed 23 November 2015.

28 Directive on the Legal Review on Weapons, Methods and Means of Warfare, established by the Norwegian Ministry of Defence, 18 June 2003.

29 The FFU consists of representatives for the following organizations: (i) the section for Operative Planning in the Department of Operational and Emergency Response Planning in the Norwegian Ministry of Defence (ii) the Norwegian Joint Operative Headquarters, (iii) The Norwegian Defence Staff College (iv) The Norwegian Defence Logistical Organization, and (v) the Norwegian Defence Research Institute. If necessary, the committee may be reinforced by other experts, like experts on medical, weapons technical qualities, or from entities outside the Norwegian military forces, if not precluded by serious concerns.

30 Directive on the Legal Review on Weapons, Methods and Means of Warfare, established by the Norwegian Ministry of Defence, 18 June 2003. This directive has replaced and repealed the Ministry of Defence Directive of 2 November 1998.

31 Daoust Isabelle, Robin Coupland and Rikke Ishoey, New Wars, New Weapons: The Obligation of States to Assess the Legality of Means and Methods of Warfare, *International Review of the Red Cross*, Vol. 84, No. 846, June 2002, p. 345-362.

32 The Joint Service Manual of the Law of Armed Conflict, JSP 383 (2004), para 6.20.1.

33 The legal review process under Article 36 consider certain key questions related to a weapon, namely: (i) Whether it is prohibited by any specific treaty provision; (ii) Whether it is of a nature to cause unnecessary suffering or superfluous injury; (iii) Whether it is capable of being used discriminately; (iv) Whether it will cause long-term, widespread and severe damage to the natural environment; (v) Current and possible future trends in IHL.

34   The British MoD has recently published the result of the legal review of its
      depleted uranium (DU) anti-armour tank rounds, known as CHARM3.
      The review committee reported that in view of the scientific studies, and the
      continuing military imperative underpinning the retention of CHARM3 as a
      weapon system, its use does not offend the principle prohibiting superfluous
      injury or unnecessary suffering in armed conflict. It also held that the
      crew training, weapon design and automated targeting systems indicate
      that CHARM3 is capable of being used discriminately. Further, where
      DU ordnance residues have existed, in the aftermath of an armed conflict,
      annual potential radiation doses have been shown by scientific study to be
      well below the annual doses received by the general population from sources
      of natural radiation in the environment and far below the reference level
      recommended by the International Atomic Energy Agency as a criterion
      to determine whether remedial action is necessary. The weapon review
      committee concluded that CHARM3 was capable of being used lawfully by
      the UK Armed Forces in an international armed conflict. Since the 1991 Gulf
      War, concern over the health and environmental effects of depleted uranium
      (DU) weapons has continued to grow. It appears that the UK has downplayed
      the potential health risks posed by exposure to DU. The Review report of 12
      July 2012, available at: http://www.bandepleteduranium.org/en/docs/192.pdf,
      accessed 20 August 2015.

35   The British Joint Service Manual of the Law of Armed Conflict, JSP 383 (2004),
      para 6.17 provides, "There is no specific rule of international law, express or
      implied, which prohibits the use of nuclear weapons. The rules introduced
      by the 1977 Additional Protocol I apply exclusively to conventional weapons
      without prejudice to any other rules of international law applicable to other
      types of weapons. In particular, the rules so introduced do not have any effect
      on and do not regulate or prohibit the use of nuclear weapons."

36   In the US, each of the military services (army, navy and air force) has issued
      instructions for the review of weapons. Since the Marine Corps and navy are
      two separate services within the Department of the Navy, the Judge Advocate
      General of the Navy, in collaboration with the Staff Judge Advocate to the
      Commandant of the Marine Corps, conduct the legal reviews for all weapons
      acquired by the Marine Corps and the navy. Department of the Army, 'Review
      of legality of weapons under international law', Army Regulation 27-53, 1 Jan.
      1979; Department of the Navy, Secretary of the Navy, 'Department of the Navy
      implementation and operation of the defense acquisition system and the joint
      capabilities integration and development system', Instruction 5000.2E, 1 Sep.
      2011; and Department of the Air Force, 'Legal reviews of weapons and cyber
      capabilities', Instruction 51-402, 27 July 2011. Paragraph 6.2, the US DoD Law
      of War Manual (June 2015).

37   Paragraph 6.4, the US DoD Law of War Manual (June 2015).

38 The DoD Manual paragraph 6.3.1.1 provides that in some cases military orders may restrict the use of certain weapons to only certain authorized purposes; there is no rule of IHL that requires that weapons or tools only be used for the purposes for which they were designed. It supports the statement with an example that an entrenching tool may be designed for digging fighting positions. However, its use as a weapon is not prohibited by IHL. Similarly, it is not prohibited to use a laser that is not designed to blind enemy persons as a weapon to blind an attacking enemy; or is not prohibited to use a weapon that has been designed to destroy enemy material, such as a large-caliber machine gun, against enemy personnel.

39 Autonomy in Weapon Systems, The US Department of Defence Directive (Number 3000.09), 21 November 2012, Part II, definitions.

40 Autonomy in Weapon Systems, The US Department of Defence Directive (Number 3000.09), 21 November 2012, enclosure 3.

41 Article 57 of the 1977 AP I states that, "In the conduct of military operations, constant care shall be taken to spare civilian population, civilians and civilian object." Further, the party responsible for an attack "shall take all feasible precautions in the choice of means and methods of attack with a view to avoiding, and in any event to minimizing, incidental loss of civilian life, injury to civilians and damage to civilian property.

42 The term 'unintended engagement' has been defined as: The use of force resulting in damage to persons or objects that human operators did not intend to be the targets of US military operations, including unacceptable levels of collateral damage beyond those consistent with the law of war, ROE, and commander's intent. Autonomy in Weapon Systems, The US Department of Defence Directive (Number 3000.09), 21 November 2012, para 1 (b) and part II, definitions.

43 Lawand Kathleen *et al.*, *Guide to the Legal Review of New Weapons, Means and Methods of Warfare: Measures to Implement Article 36 of Additional Protocol I*, Geneva: ICRC, 2006, p. 4.

44 Schmitt, Michael N., Autonomous Weapon Systems and International Humanitarian Law: A Reply to the Critics, *Harvard National Security Journal Features*, 2013, p. 28.

45 The ICRC has called on States to establish mechanisms and procedures to determine the conformity of weapons with international law. In particular, the 28th Conference declared: "In the light of the rapid development of weapons technology and in order to protect civilians from the indiscriminate effects of weapons and combatants from unnecessary suffering and prohibited weapons, all new weapons, means and methods of warfare should be subject to rigorous and multidisciplinary review." The International Conference of the

Red Cross and Red Crescent in the 27th (1999) and the 28th Conference (2003) called on the States to establish mechanisms and procedures to determine the conformity of weapons with international law. Final Goal 2.5 of the Agenda for Humanitarian Action adopted by the 28th International Conference of the Red Cross and Red Crescent (2003).

46 For example, the UK government's reservation to AP I states that "the rules so introduced do not have any effect on and do not regulate or prohibit the use of nuclear weapons."

47 Boulanin Vincent, Implementing Article 36 Weapon Reviews in the Light of Increasing Autonomy in Weapon Systems, SIPRI Insight on Peace and Security, No. 2015/1, November 2015, p. 4.

48 Robert Nash is the director of "Article 36" and joint Coordinator of the International Network on Explosive Weapons.

49 Copeland Damian P., 'Legal Review of New Technology Weapon', in Nasu Hitoshi and Robert McLaughlin (ed.). 2014. *New Technologies and the Law of Armed Conflict*, The Hague: TMC Asser Press, p. 43-55.

50 Noel Sharkey, Cassandra or false prophet of doom: AI robots and war, in IEEE Intelligent Systems, Vol. 23, 2008, p. 17.

51 The Martens Clause states: "In cases not covered by this Protocol or by any other international agreements, civilians and combatants remain under the protection and authority of the principles of international law derived from established custom, from the principles of humanity and from dictates of public conscience." Also see: Weizmann Nathalie, Autonomous Weapon Systems under International Law, Academy Briefing No. 8, Geneva Academy of International Humanitarian Law and Human Rights, November 2014, p. 17-20.

52 International Court of Justice (ICJ), Legality of the Threat or Use of Nuclear Weapons, Advisory Opinion, ICJ Reports, 1996, para. 78.

53 Brown Gary D. and Metcalf Andrew, Easier Said Than Done: Legal Reviews of Cyber Weapons, *Journal of National Security Law and Policy*, Vol. 7, 2014, p. 115-138.

54 Boulanin Vincent, Implementing Article 36 Weapon Reviews in the Light of Increasing Autonomy in Weapon Systems, SIPRI Insight on Peace and Security, No. 2015/1, November 2015.

55 In September 2015, the Stockholm International Peace Research Institute (SIPRI) convened an expert seminar in Stockholm. The objective of the seminar was to provide a platform for interested states parties to share their

experience of conducting legal reviews of weapons. The participating states were France, Germany, Sweden, Switzerland, the UK and the USA. The discussions were facilitated by experts from the ICRC, SIPRI, the Swedish Defence Research Agency, the Swedish Defence University and the Swedish Red Cross. For more details on the issue see: Boulanin Vincent, Implementing Article 36 Weapon Reviews in the Light of Increasing Autonomy in Weapon Systems, SIPRI Insight on Peace and Security, No. 2015/1, November 2015.

56   According to Boothby (2009), this is the logical interpretation of the stipulation in Article 3 of the Hague Convention IV, 1907, that states are responsible for all acts of the armed forces. Boothby William H. 2009. *Weapons and the Law of Armed Conflict*, Oxford: Oxford University Press, p. 341.

57   *A Guide to the Legal Review of New Weapons, Means and Methods of Warfare: Measures to Implement Article 36 of Additional Protocol I of 1977.* 2006. Geneva: ICRC, p. 4. In 2006, in order to assist States, the ICRC prepared the Guide in consultation with military and international law experts from ten countries. Available at: http://www.icrc.org/eng/assets/files/other/icrc_002_0902.pdf, accessed 15 March 2011.

58   Coupland Robin M., The SIRUS Project: Towards a Determination of which Weapon cause "Superfluous Injury or Unnecessary Suffering", Geneva: ICRC, 1997.

59   As on 31 December 2015, the 1977 Additional Protocol I has a total of 174 states parties. It has not been ratified by a few States like India, Iran, Israel, Malaysia, Pakistan, Turkey and the US.

60   Jacobsson Marie, Modern Weaponry and Warfare: The Application of Article 36 of Additional Protocol I by Governments, *International Law Studies*, Vol. 82, 2006, p. 183-191.

61   Boothby William H. 2014. 'The Legal Challenges of New Technologies: An Overview', in Nasu H. and R. McLaughlin (eds.), *New Technologies and the Law of Armed Conflict*, TMC Asser Press, p. 21-28.

VII LAWS: International Concerns

Introduction

In the last five years, the international community has raised questions related to the legal, technical and ethical consequences of the use of lethal autonomous weapons systems (LAWS). In May 2013, Prof Christof Heyns, the United Nations (UN) special rapporteur on extrajudicial killings, submitted a report to the UN Human Rights Council that raised many objections to this emerging robotic technology and called for national moratoria on the production, transfer, acquisition, and use of such technology. Six months later, the State Parties to the Convention on Conventional Weapons (CCW)[1] debated issues related to LAWS.[2] This was followed by an informal meeting of experts in May 2014 under the auspices of the CCW during which representatives from 87 States talked about the future of LAWS. In April 2015, a meeting of experts at Geneva also discussed emerging technologies related to LAWS in the context of the objectives and purposes of the CCW. International concern regarding LAWS has also been expressed in other UN forums and in some reports of international NGOs like the International Committee of the Red Cross (ICRC). Some of these views will be discussed in this chapter.[3]

Report of UN Special Rapporteur

Prof Christof Heyns, the UN Special Rapporteur on extrajudicial, summary or arbitrary executions,[4] is of the firm opinion that the induction of LAWS into the arsenals of the States would add a new dimension to the conduct of warfare as targeting decisions could be taken by the robots themselves. In addition to being physically removed from the kinetic action, people would also become more detached from the decisions to kill an individual.

There would be a risk of the distinction between weapons and warriors becoming blurred, as the former would take autonomous decisions about their own use.

Heyns pointed out that one of the most difficult issues that the legal, moral and religious codes of the world have grappled with is the killing of one human being by another. The prospect of a future in which LAWS could exercise the power of life and death over human beings would complicate the issue further. LAWS have the potential to pose new threats to the right to life under the IHL as well as international human rights law. They could also create serious international division and weaken the role and rule of international law, and in the process, undermine the international security system. Thus, before incorporating them into our arsenals, all involved i.e. States, international organizations and civil societies, must consider the full implication of embarking on this road. There is a need to remind ourselves of the supremacy and non-derogability of the right to life under both treaty and customary international law.

The 2014 CCW Meeting of Experts on LAWS

During the first informal CCW Meeting of Experts on LAWS in Geneva from 13 to 16 May 2014, no participating State admitted to developing LAWS. However, many of them stressed that the notion of meaningful human control (MHC) could be useful in addressing the issue of autonomy.[5] A few States stated explicitly that weapon systems lacking MHC were unacceptable. Although this concept has not yet been defined and more work is needed to understand whether and how it might be operationalized, its resonance in international discourse suggests that it could be indicative of an emerging norm. The views expressed by some of the Member States during the meeting and on earlier occasions are as follows.

Austria

Austria was of the firm view that any use of a weapon in armed conflict must comply with international humanitarian law norms. It stated that among roboticists and lawyers alike, there is serious doubt that autonomous weapons can ever be programmed in a way to guarantee this compliance. Further, while in the case of a war crime perpetrated by a human actor legal responsibility can be, at least in principle, established, it is fundamentally unclear as to how to resolve the issue when the autonomous decision of a machine is at the root of the crime. Austria made a plea for great caution to be exercised before pushing forward technological developments the implications of which seem not to be sufficiently understood yet.

Brazil

The permanent representative of Brazil asserted that permanent, universal and inclusive decisions on international issues could be reached only through multilateral means and the CCW was an appropriate forum for discussing any future multilateral regime regarding the regulation of LAWS. Taking into account the rapid pace of scientific developments in this sphere, it was critical to ensure that all new and emerging technologies are employed in the military field in conformity with human rights and IHL. The ambassador was of the view that the Martens Clause was a keystone of IHL and was hopeful that the discussions during the informal meeting would be useful in identifying ways of reconciling the military perspective on the use of LAWS with the principles of IHL, human rights and ethical and moral imperatives.

Canada

Canada observed that the existing IHL framework was sufficient to regulate the use of LAWS, but acknowledged that there was a compelling need to regulate these weapons. It also expressed interest in discussing the notion of MHC over LAWS.

China

China expressed support for the CCW efforts and was of the view that more information and understanding was required with respect to humanitarian concerns, and legal and military issues, including definition, scope, and applicability of the existing laws to LAWS. It was in favour of further discussions to improve understanding and build consensus among the State Parties.

Croatia

The Republic of Croatia asserted that the CCW had always been seen as a forum for open-minded discussions and exchange of views on disarmament issues at the multilateral level. It was of the opinion that while it is recognized that humans must retain ultimate control, more detailed deliberations are needed on what constitutes adequate, meaningful, or appropriate human control over the use of force. It hoped that the CCW talks would eventually lead to a legally binding instrument and usher in a new era of hope for the international disarmament community.

Czech Republic

According to the Czech Republic, if it has been rather difficult to strike a balance between humanitarian concerns and security requirements in the past, it will even more challenging to do so in the case of autonomous weapons of the future. Hence, it is important to start working on ways to protect civilians and combatants from the possible effects of LAWS before they are developed.

Egypt

Egypt was of the view that the informal meeting of experts on LAWS could serve as an eye-opener on a very important and challenging aspect of weaponry research and development. It felt that the possible impact of LAWS on the value of human lives and the cost of war needed to be discussed, together with the possibility of the acquisition of such weapons by terrorist and organized crime networks. Such discussions ought to lead to the prohibition of acquisition, research and development, testing, deployment, transfer and use of LAWS, and until such a result was achieved, it would support calls to pose a moratorium on the development of LAWS. As military robotics gain more and more autonomy, the ethical questions involved would become even more complex and it might be too late to work on an appropriate response.

France

France has actively encouraged a substantive debate on the technical, legal, ethical, and operational aspects of LAWS. In May 2013, at the Human Rights Council, it stated that it does not possess and does not intend to acquire robotized weapons systems with the capacity to fire independently. It believes that military and political leaders must be fully responsible for any decision to use armed force and that the role of human beings in the decision to open fire must be preserved. France has been maintaining that it is essential to maintain human control in the operational loop of military robots, especially weaponized robots. The act of war, which results in delivering death, must involve substantial human responsibility, both at the level of political leadership and the military chain of command. For ethical and operational reasons, the human conscience cannot be taken out of the equation. However, it is necessary to bear in mind that any technology is of a dual nature, and since robotics may have many civil, peaceful, legitimate and useful applications, there can be no question of limiting research in this field.

Germany

Germany too was of the opinion that there must be a common understanding in the international community that it is indispensable to maintain human control over the decision to kill another human being. It stated that the principle of human control was the foundation of international humanitarian law and that even in times of war, human beings could not be made objects of machine action. It declared that it did not intend to have any weapon system that takes away human control over the decision about life and death.

Germany was doubtful whether LAWS would be able to discriminate between combatants as legitimate targets and non-combatants; or apply the principle of proportionality to assess whether the possible collateral damages could be justified by the achievable military advantage. According to Germany, many experts and scientists were doubtful whether a computer would ever be able to make such qualitative evaluations. Germany was against the use of weapon systems without human control and stated that the international community should arrive at a consensus about this topic.

Holy See

According to Archbishop Silvano M. Tomasi, permanent representative of the Holy See to the United Nations, the fundamental question was whether even well-programmed machines with highly sophisticated algorithms to make decisions that seek to comply with IHL could truly replace humans in decisions over life and death. Humans, he felt, must not be taken out of the loop over decisions regarding life and death. Meaningful human intervention over such decisions must always be present.

India

The Permanent Mission of India was non-committal on the issue of LAWS. He said a fresh look was required on whether LAWS meet the criteria of international law and IHL, especially with regard to the principles of distinction, proportionality and precaution; and suggested a preemptive ban on research, production and use of LAWS until there was greater clarity on the overall implications. He also said that there was a spectrum of autonomy built into the existing weapon systems and that a ban on LAWS would be premature, unnecessary and unenforceable. India has consistently urged for a more broad-based discussion on the impact of the proliferation of LAWS on international security.

The Indian delegation was of the opinion that there was a need for measures to control the widening of the technology gap among States so as to discourage the use of lethal force to settle international disputes since it would cause fewer casualties on one side. It highlighted the significance of the Martens Clause as an important reference point, but said that it may not be an adequate filter for the development of new weapons contrary to IHL. India also raised the issue of whether Article 36 (of 1977 Additional Protocol I) reviews of LAWS would suffice since the language of this provision was developed when human role was central to the use of force. India warned against rushing to a decision on MHC as this might lead to the legitimization of weapons.

Ireland

Ireland stressed the importance of effective human control over the use of force. It supported the CCW's mandate to regulate or ban the use of specific categories of conventional weapons that have effects which are unacceptable to the conscience of humanity. Ireland stated that the debate over lethal autonomous weapons went far beyond legal and technical complexities, raising fundamental questions about the role of humans in taking lethal decisions in armed conflict. It also expressed concern over the potential use of LAWS in law enforcement and other situations beyond the scope of the CCW mandate.

Israel

At the experts meeting in May 2014, Israel exhorted delegates to have an open mind regarding LAWS, as it was difficult to foresee how they would develop over the next 10-30 years. It maintained that it was difficult to assert at this stage that LAWS could never reach certain capabilities in the near future. It proposed that LAWS be assessed on a case-by-case basis and said that each system must be adapted to the complexity of the environment of use, which could be simplified by limiting system operations for specific territory, targets, tasks, or other limitations set by a human. The system could, if necessary, be programmed to refrain from action and await input from a human when the situation is unclear. It contended that LAWS might comply better with IHL because they would be more predictable and unemotional. It did not address the concept of MHC at the CCW's annual meeting in November 2014, but maintained that human judgment comes into play throughout the various stages of development, testing, review, approval, and decision to employ a weapon system, including an autonomous one.

Italy

Italy stated that the CCW had the potential to address not only humanitarian concerns over existing weapons but also to prevent the development of new types of weapons that would be unacceptable under basic IHL. It was hopeful that the potential threats envisaged from LAWS could be discussed by the CCW.

Japan

At the May 2014 experts meeting, Japan was of the view that there was no room to question whether LAWS could comply with IHL. It declared that it had no plan to develop lethal weapon systems without humans in the loop. In November 2014, it stated that it was imperative to develop a common understanding about what we perceive as LAWS in order to advance discussions. It said, "While we may continue researching and developing non-lethal autonomous technology for defence purposes, we are not convinced of the need to develop fully lethal autonomous weapon systems which are completely out control of human intervention."

Mexico

Mexico was of the opinion that LAWS could not comply with the key principles of IHL and there was a need for a significant level of human control in addition to the legal review of new weapons under Article 36. It warned that LAWS could be used in situations other than armed conflict, so it would be necessary to consider international human rights law as well as IHL. At the annual meeting in November 2014, Mexico stated that the Martens Clause could be an effective tool to address the issue of the rapid evolution of military technology and corresponding customary international law.

The Netherlands

The Netherlands has been of the view that LAWS are probably not inherently illegal and that their predictability is the main issue. At the 2014 experts meeting, it said that its committee of experts rigorously tests and reviews new methods and means of warfare in what is not solely a legal review, but a process that involves ethical, technical, social, and political considerations. It was of the opinion that the core issue concerning LAWS was that once activated, it could select and engage targets without human supervision or intervention. It urged further substantive discussion on several issues relating to such weapons, particularly on meaningful human control, ethical aspects, and weapons reviews. The Netherlands has been supporting research by the UN Institute for Disarmament Research (UNIDIR) on autonomy in weapons systems.

Norway

Norway questioned whether LAWS could be programmed to undertake a complicated analysis to make a proportionality assessment without human intervention. It defined LAWS as "weapon systems that search for, identify and use lethal force to attack targets, including human beings, without a human operator intervening, and without meaningful human control." It said that the possible development of such weapon systems raised a number of ethical and legal questions and felt that it was a necessity to ensure that the basic rules and principles of international law were upheld.

Pakistan

The delegation from Pakistan firmly believed that the development and deployment of LAWS would have a wide range of implications, not just in the field of disarmament but also under international human rights and IHL. It felt that a situation in which a party to a conflict bears only economic costs and its combatants are not exposed to any danger, is not war but one-sided killing. LAWS would have the same problems that lethal drones have. In addition, the development of such weapon systems would impact the developing countries disproportionately; therefore, there was a need to move beyond national moratoria. The international community should consider a ban on the use of LAWS as in the case of blinding laser weapons under CCW Protocol IV.

The Republic of Korea

The delegate from the Republic of Korea said that there was a need to consider the effects of LAWS on future armed conflicts, and assess the challenges posed by them to IHL. The issue must be discussed in a focused manner within the framework of the CCW, to which almost all countries possessing significant robot technology are States Parties. The discussions, however, should not be carried out in a way that would lead to restrictions on the development and manufacture of non-lethal robots, in particular industrial robots used for peaceful purposes. The delegate also said that it was important to reach a common understanding on the relevant terminology, which is both complex and less familiar to many members of the disarmament community.

The Republic of Korea is moving forward with efforts to enhance the use of robots in various industries. In the military arena, it is considering the utilization of robot technology to better protect soldiers exposed to serious risks, for instance, by using robots to clear mines, remove improvised explosive devices (IEDs), and detect CBRN.[6]

Russia

Russia expressed concern over the implications of the use of LAWS for societal foundations, including the negation of human life. It feared that such machines could also significantly undermine the ability of the international legal system to maintain minimal legal order. The Russian delegate stressed the need to ensure transparency in all aspects of the development of LAWS and to take into account the standards of IHL and international human rights law at all stages of the development of lethal robotics.

Sierra Leone

According to Sierra Leone, it was time to view robotic technology under the human rights lens. Though a robotic weapon can be programmed to minimize errors and reach their targets with a high degree of accuracy, robots are machines and can kill innocent people indiscriminately. Further, as in the case of all other technology, LAWS can fall into the wrong hands and be used irresponsibly. The delegation was of the firm view that the Human Rights Council should call on all States to declare and implement national moratoria on the testing, production, assembly, transfer, acquisition, deployment and use of LAWS until such time as an internationally agreed-upon framework on the future of such weapons was established.

South Africa

The Government of South Africa stated that the development of LAWS poses serious questions and there were many issues on which clarity was required. This includes definitional certainty as to the notion of autonomous and semi-autonomous weapons systems. The primary concern of the South African delegation was the humanitarian implications of the use of LAWS and related ethical considerations; and whether these new technologies of warfare would be compliant with the rules of IHL, including those of distinction, proportionality and military necessity.

Sri Lanka

The Sri Lankan delegates felt there were many areas that needed to be fully analysed, debated and understood, including principles and concepts, implications, and the types of action required. They were of the view that the success achieved in prohibiting blinding laser weapons by adopting Protocol IV to the CCW, was a pointer to the possibility of effectively addressing this challenge.

Sri Lanka believes that due consideration should be given to the potential benefits of the peaceful use of robotic technology such as in rescue operations, intelligence, mine clearance, logistical operations, as well as other areas such as agriculture and health. There must be absolute clarity among State Parties on certain areas of the potential use of LAWS even in the limited circumstances that might be identified for the purpose. These areas include the level of autonomy, the context in which such technology is used, the potential risks, fixing accountability and compliance with IHL and other laws, and ethical and sociological obligations.

The Sri Lankan delegates were particularly concerned about the possible impact of the use of LAWS on international peace and security, as such autonomous systems have the potential to escalate the pace of warfare and undermine the existing arms controls and regulations, to aggravate the dangers of asymmetric warfare, and destabilize regional and global security. The possession of autonomous weapons by some States and the possibility of the asymmetric use of these weapons in war may compel other States to abandon their policies of restraint or moratorium and ignite an arms race. The experience with nuclear weapons, they pointed out, provides a useful lesson on the consequences of such an arm race.

The delegates urged the international community to think seriously about the ethical question of leaving the ultimate decision on the right to life to a machine that operates without human control. They felt that we need to be wary of allowing any level of autonomy in the use of weapons systems.

Sweden

At the experts meeting in May 2014, Sweden held that while it was true that many systems with various degrees of automation are being used or developed by states, it was not clear if this amounted to a move toward systems that would give full combat autonomy to machines. Fixing the threshold at which a weapon should be considered 'autonomous' would be a difficult task because automation/autonomy exists on a continuum. However, an autonomous weapon that is fully outside the control of a human was not a desirable development for a military force.

Switzerland

Switzerland[7] stated that no one wants to see a battlefield with machines entrusted with deciding who lives and dies. At the annual meeting in November 2014, Switzerland proposed that the concept of MHC be considered in greater detail.

United Kingdom

In a 2011 Joint Doctrine Notes, UK's Ministry of Defence stated that it had no intention to develop systems that operate without human intervention in the weapon command and control chain. However, it was looking to increase levels of automation to make systems more effective. The British delegate stated that existing international law was sufficient to regulate the use of LAWS; therefore Britain had no plans to call for or support an international ban on LAWS.

United States

The US policy on autonomy in weapon systems is contained in the Department of Defense Directive of November 2012.[8] It covers manned and unmanned platforms, as well as guided munitions, and excludes mines, cyber weapons, and manually guided munitions. In May 2013, at the Human Rights Council, the US was in agreement that LAWS may present important legal, policy, and ethical issues. It called on all the States to proceed in a lawful, prudent, and responsible manner when considering whether to incorporate automated and autonomous capabilities in weapons systems. During the CCW's annual meeting in November 2014, the US proposed a focus on the technical, legal, operational policy-related challenges in the sphere of autonomy. It said that the weapons review process could provide the basis for identifying fundamental issues and provide guidance for States that are considering any new weapons system and that a discussion could result in a set of best practices applicable to the future development of LAWS. The US was of the belief that a flexible and responsible framework for the development and use of autonomous capabilities in weapons systems must be established. It must include a stringent review process for new autonomous weapons likely to be developed in the future.

International Committee of the Red Cross (ICRC)

The ICRC has shown concern regarding LAWS since 2011.[9] It has called upon the States to carefully consider the fundamental legal, ethical and societal issues raised by these weapons before developing and deploying them. In November 2013, the ICRC opined that at this juncture, it was not joining calls for a moratorium or ban on LAWS, and asked the States to fully consider the fundamental legal, ethical and societal issues related to the use of these weapons well before they are developed, and to ensure that such weapons are not used if there is no certainty that they can comply with IHL. The ICRC further stated that though there was a wealth expert literature on this subject, there was a lack of consistency in the use of terms. Should States decide to discuss autonomous weapons in a more focused way in either the CCW or elsewhere, there would be a need to look more closely at terms and definitions.

The ICRC explained that automated weapons function in a self-contained and independent manner although they may be initially deployed or directed by a human operator. Once activated, these systems can engage individual targets or specific "target groups" that have been selected (programmed) by a human operator. They execute precisely preprogrammed actions or sequences within a well-defined and controlled environment. "Our understanding is that these highly automated systems in practice operate with a 'human on the loop', i.e. under human supervision.

The ICRC has expressed a number of concerns regarding the capability of LAWS; in particular, developing the capacity of such weapon systems to fully comply with the IHL rules of distinction, proportionality and precautions in attack. The ICRC is of the opinion that in future it may technologically be possible to programme an autonomous weapon system to fully comply with the rules of IHL. Such an autonomous weapon may be able behave more ethically and more cautiously on the battlefield than a human being – taking for granted that a robot would not be affected by emotion or personal self-interest.

Another key concern expressed by the ICRC is the issue of accountability for violations of IHL committed by LAWS. State responsibility and individual criminal liability for serious violations of IHL (war crimes) are essential elements for the protection of victims of armed conflict. If the act can be attributable to a particular State, that State would be responsible for compensation under both IHL and the rules of State responsibility.

But it would be extremely difficult to hold an individual accountable for a decision made by LAWS, and if responsibility cannot be determined as required by IHL, it would unethical to deploy such autonomous systems.

The ICRC in its subsequent meeting of experts held at Geneva in March 2014,[10] held that there was no doubt that the development and use of LAWS in armed conflict is governed by IHL, including the obligation to undertake legal reviews in the study, development, acquisition or adoption of new weapons. The legality of LAWS must be assessed on the basis of their design-dependent effects and their intended use. The ICRC was doubtful whether such weapons could be adequately tested given the absence of standard methods for testing and evaluating autonomous systems.

Some experts at the meeting felt that there was little interest in deploying LAWS to replace humans in an operational context, given the necessity to keep control over the operations from the commander's point of view. The relevance of using autonomous technologies for tasks, such as intelligence gathering, rescue, protection of armed forces and civilians, logistics and transportation, was discussed. The risks associated with the use of LAWS, including vulnerability to cyber attacks, lack of predictability and difficulties of adaptation to a complex environment, were stressed by some delegations. The question of interoperability with allied forces was mentioned and the capacity of LAWS to comply with the rules of international law was discussed. A few experts underlined the need to consider the environment in which the system would operate (air, land or sea) to determine the military relevance of LAWS and to carry out a proper risk assessment.

The 2015 CCW Meeting of Experts

The CCW Meeting of Experts on LAWS was held from 13 to 17 April 2015 at the United Nations in Geneva. [11] It was a mega event attended by representatives of the State parties[12], members of the civil society, non-governmental organizations,[13] and experts in law, computer technology and ethics. Representatives of the United Nations Institute for Disarmament Research (UNIDIR), United Nations Interregional Crime and Justice Research Institute (UNICRI), United Nations Office of Disarmament Affairs (UNODA), European Union, International Committee of the Red Cross (ICRC) and Geneva International Centre for Humanitarian Demining (GICHD),[14] also participated in the meeting.

The Chair-Designate of the Convention, Sri Lankan Ambassador Aryasinha remarked that LAWS have been described as the next revolution in military technology, and that issues related to the development, production, and use of fully autonomous weapons require serious consideration. Aryasinha felt that the CCW provides a flexible framework for debates on LAWS by the State parties and other stakeholders. Views expressed by the delegates, experts and non-governmental organizations are as follows.[15]

Austria

In a working paper submitted at the CCW meeting, Austria highlighted that the prospect of weapons taking decisions about the use of force without human intervention poses a challenge to IHL. It said the concept of 'meaningful human control' should not be seen as a new legal norm, but as evaluating LAWS on the basis of the existing standards of IHL. New weapons need to comply with the three fundamental IHL principles, namely the principle of proportionality, distinction and precaution. The lawfulness of a new weapon should be judged in accordance with Article 36, which stipulates the obligation of every State Party to determine whether the employment of the weapon would, in some or all circumstances, be prohibited by the Protocol or by any other rule of international law applicable to the High Contracting Party.

Chile

The Permanent Representative of Chile to the United Nations was of the opinion that the proportionality principle of IHL may be placed in jeopardy with the use of lethal force by LAWS. He was of the view that the Martens Clause is applicable to LAWS, as are all subsequent legal and political developments of IHL and human rights law. The International Court of Justice in 1996, in its advisory opinion on the legality of the threat or use of nuclear weapons, referred to the Martens Clause stating that it has proved to be an effective means of addressing the rapid evolution of military technology. This is a valid criterion that should necessarily be applied to an emerging technology the consequences of which are hard to predict, although it would need to be considered in varying degrees with regard to nuclear weapons and to this new type of weapons. It has become clearer that the battlefield use of LAWS would potentially affect human rights, including the right to life, the right to dignity, the right to freedom and security and the prohibition of torture and other forms of cruel, inhumane or degrading treatment. Both branches of international law—IHL and human rights—are to be taken into consideration for the protection of human beings and the protection offered by human rights conventions does not cease in case of armed conflict.

Czech Republic

The Czech representative was of the view that although LAWS do not yet exist, systems equipped with increasingly advanced autonomous features/ capabilities have been introduced and are operated by a number of states including his country. This is an evolution that cannot be reversed, be it in the military or civilian sector. From the humanitarian point of view it might be more reasonable to concentrate on certain critical autonomous features of weapons that could be regulated or prohibited, rather than to pursue an absolute ban on these weapons. He felt that the ultimate decision to end somebody's life must remain under MHC and this principle, inherent in IHL, should be respected by the international community. The Czech Republic, he said, was fully aware of the implications of the introduction of weapons with autonomous capabilities and was ready to work on provisions to help minimize unintended and unacceptable engagements.

Germany

Germany reiterated its position on LAWS, stating that it does not support the view that the decision to use force, in particular the decision over life and death, may be taken by an autonomous system without any human intervention in the selection and engagement of targets. The German representative said that it was for good reasons that the fundamental principles of IHL mostly avoid specifying particular technologies. They are applicable irrespective of the type of weapon, and means and methods of warfare, and define the difference between legitimate and illegitimate use by focusing on the effects of deployment. Any possible understanding in respect of LAWS will have to comply with these principles.

India

The Indian delegate was of the view that forum of the CCW should be strengthened in terms of its objectives and purposes. This should be achieved through increased systemic controls on international armed conflict in a manner that does not widen the technology gap amongst states or encourages the increased resort to military force in the expectation of lesser casualties. The delegate also said that there was wide divergence on issues such as MHC in relation to LAWS and it was not clear whether distinction could be drawn between oversight, review, control or judgment or how they would apply to new weapon system. He pointed out that the inexorable march of new technology, and the dual nature of autonomous systems could not be ignored.

Ireland

Ireland's position was that the weapons should remain under effective human control. It expressed concern that LAWS could be used outside of traditional combat situations, for example in law enforcement. It therefore stressed the relevance of discussing such weapons under international human rights law as well as IHL.

Israel

The representative of Israel was of the opinion that deliberations on LAWS should be guided by three basic assumptions: (1) The necessity to maintain an open mind regarding both the potential risks and the possible positive capabilities of future LAWS. Since it was difficult to foresee how autonomous capabilities may develop in the next 10 to 50 years, any discussion on future LAWS should be undertaken in a cautious and prudent fashion. (2) An assessment of such systems and of their employment should be conducted on a case-by-case basis. (3) The use of future LAWS, as of any other means of warfare, must comply with the applicable rules of IHL. The States should, when developing LAWS, subject the system to an internal legal review.

Japan

Japan's position was: (1) We need to be well-informed and assess the current state of affairs with regard to the research and development of autonomous functions in technical systems and the future trends in robotics. With a proper understanding of the current situation, we can tackle future challenges in a more effective manner. (2) It is necessary to clarify the definition of LAWS, for which we need to conduct in-depth discussions on the main elements of LAWS, such as autonomy and meaningful human control. (3) It is necessary to keep in mind the dual-use nature of robotic technology, which is being used extensively in industries, disaster response, and health care. It may be difficult to draw a line between technical components enabling autonomy in civil applications and those in military use. The delegates declared that Japan had no plan to develop lethal robots without humans in the loop.

Pakistan

Pakistan claimed that its position on LAWS was consistent, clear and unambiguous. It firmly believes that LAWS by their very nature would be unethical, as they would be incapable of distinguishing between combatants and non-combatants. With the prospect of no loss or injury to their "human" combatants, States employing LAWS would resort to the use of force frequently and undermine the very basis of the restraints on the use of force that international law seeks to maintain. The introduction of LAWS would also affect progress on disarmament and non-proliferation.

LAWS could also risk the lives of civilians and non-combatants on both sides. It remains unclear as to how "combatants" will be defined in the case of LAWS. Will targets be chosen on the basis of an algorithm that recognizes certain physical characteristics, for example, "beards and turbans"? The use of LAWS would also create an accountability and transparency vacuum and provide impunity to the user due to the inability to attribute responsibility for the harm that they cause. If the nature of a weapon renders responsibility for its consequences impossible, it's use should be considered unethical and unlawful.

The Pakistani delegate asserted that we should not let the blind quest for the ultimate weapon, driven by the interests of the military-industry complex, get the better of us. The introduction of LAWS would be illegal, unethical, inhumane and destabilizing for international peace and security. Therefore, their further development and use must be banned through a dedicated Protocol of the CCW. Pending the negotiations and conclusions of a legally binding Protocol, the States currently developing such weapons should place an immediate moratorium on their production and use. Pakistan recommended the establishment of a Group of Governmental Experts (GGE) with a mandate to formally consider the issue of LAWS and present a report to the CCW Review Conference in 2016.

Poland

The delegation from Poland asserted that a state should always be held accountable for what it does, especially for the responsible use of weapons by its armed forces. The responsibility of States in the case of LAWS should include their development, production, acquisition, handling, storage and international transfers.

Sri Lanka

Sri Lanka said that when seeking to regulate a dual-use technology, such as robotic technology, due consideration should be given to its potential benefits in peaceful use. It was of the view that there must be maximum clarity among States Parties on several key areas related to the potential use of LAWS. The Sri Lankan delegate was concerned about the possible impact of the use of LAWS on international peace and security. Such autonomous systems have the potential to escalate the pace of warfare and undermine the existing arms controls and regulations, to aggravate the dangers of asymmetric warfare, and destabilize regional and global security. The possibility of non-state actors gaining access to LAWS could gravely endanger the security of all people, regardless of its initial intended military use. In addition, the use of LAWS could pose new challenges to compliance with IHL principles such as distinction, proportionality, precaution and military necessity.

The United Kingdom

The UK has been of the view that programming might in future give rise to an acceptable form of MHC over LAWS, and research into such possibilities should not be pre-emptively banned. The delegates of the UK emphasized that IHL provides the appropriate paradigm for discussions on new weapons and that legal review of LAWS must be undertaken at the domestic level as required under Article 36 of AP I. Further, a clear, careful and well understood definitions related to the characteristics of LAWS must be analysed under the auspices of the CCW.

The United States

The US was of the view that any use of LAWS would have to be compliant with applicable law. For purposes of law enforcement and border or crowd control, the human rights framework and national law would play a vital role in discussions about the possible future development and use of LAWS. The delegate affirmed that his country was committed to ensuring the utmost respect for and adherence to human rights when it comes to the development, use, or export of any weapons system, including any potential future LAWS. He also said that armed conflict is governed primarily by IHL and the US is committed to adherence to its principles. However, there is no specific provision in IHL that prohibits or restricts the use of autonomy to aid in the operation of weapons, including LAWS.

The US believes that a robust policy process and methodology can help mitigate risks when developing new weapon systems. It has a process in place, applicable to all weapon systems, which is designed to ensure that weapons operate safely, reliably and are understood by their human operators. The US has been following the policy of legal review of the intended acquisition of a weapon system to ensure that its development and use is consistent with applicable law, including IHL.

The US firmly believes that the Martens Clause does not lead to categorical legal prohibition of LAWS. It is not a rule of international law that prohibits any particular weapon, much less a weapon that does not currently exist. In general, the lawfulness of the use of a type of weapon under international law does not depend on an absence of authorization, but upon whether the weapon is prohibited. Given that there are significantly divergent views on the definition of LAWS, the US is of the view that such systems are not prohibited by customary law. It believes that the principles of humanity and the dictates of public conscience provide a relevant and important paradigm for discussing the moral or ethical issues related to the use of automation in warfare.

According to the US, in certain contexts, the use of weapons systems with autonomous functions may be preferable to other weapons of today, as the former are more precise and less likely to cause collateral damage. LAWS may even lower the threshold for engaging in peacekeeping operations, which could save lives, or prevent people from committing gross human rights violations in situations where "we might otherwise hesitate to put boots on the ground".

The International Committee for Robot Arms Control

The International Committee for Robot Arms Control (ICRAC) [16] is of the view that the robotics revolution can help to make our lives better by taking care of dull, dirty, and dangerous tasks, increasing productivity and helping to mitigate the impacts of natural disasters and assisting humanity in general. However, the development of LAWS would pose a host of problems for global security. Once a new weapon has been developed and creates any military advantage, other nations rush to acquire it. With no international ban on the development of LAWS, there is a likelihood of the mass proliferation of these weapons. The presence of LAWS in the military arsenal could also lead to their use in conflicts short of warfare, thus lowering the threshold for armed conflict. LAWS could be positioned, like landmines, to patrol post-conflict zones, thus creating a permanent global battlefield. Further, LAWS would remain vulnerable to cyber attack and may also fall into the hands of non-state actors.

The Centre for New American Society

Scharre (2015) provided introductory information to the UN delegates. He was of the view that though LAWS do not yet exist, the increasing autonomy in systems across militaries makes it important to discuss the issue at this juncture. He advanced three main reasons for this: (1) Rapid advances in computer technology have raised the prospect of future development of autonomous systems in many applications. (2) It is important to distinguish between trends toward greater autonomy in systems in general, such as self-driving cars, military robots, or missiles with advanced navigation features, and LAWS that would select and engage targets on their own. (3) Some simple forms of autonomous weapons already exist,[17] although they are generally limited to systems supervised by humans that protect vehicles and military bases from attacks.

On the issue of the legality of LAWS, Scharre was of the opinion that though there are no specific provisions in IHL that prohibit LAWS, like all other weapons, the use of LAWS must be compliant with IHL principles. Further, LAWS raise a number of important moral and ethical issues that are not explicitly addressed in IHL, but should be considered. For instance, during an armed conflict it is lawful to kill, but there have been many instances when humans have refrained from doing so. In theory, LAWS may not have this restraint, and their use could therefore lead to more killing in war.[18] He also said that regardless of the system's exact degree of autonomy, States should implement strict testing and evaluation standards to ensure that the system performs as intended and that the procedure for its use are clear to the operators. Systems should be tested in a realistic operating environment to ensure that they continue to operate in accordance with their design features. States should also explore the ways in which accountability for the use of LAWS can be established and enforced. While such an undertaking will be challenging, it will be critical in ensuring that the use of LAWS adheres to IHL.[19]

The Stockholm International Peace Research Institute (SIPRI)

The delegates from SIPRI highlighted that there is very little official data available outside the statements made at the CCW and only two countries (the USA and the UK) have so far issued an official policy document related to autonomous weapons. The debate on LAWS requires expertise from various ministries (e.g., defence, foreign affairs) and different departments within these ministries (e.g., the departments dealing with disarmament and issues related to international law and human rights). Interested States should appoint a LAWS focal point within their own system with the task of outreaching to all relevant agencies and authorities within their own system, to build a comprehensive picture of the current national thinking related to LAWS. They should also create a point of contact for providing other States with available information about the national approach to various dimensions of LAWS. This system should be seen as a platform sharing relevant information for the benefit of all parties to the CCW. SIPRI recommended the establishment of a voluntary but regular and systematic information sharing mechanism to support future discussion on LAWS at the CCW.

Article 36

Thomas Nash, on behalf of Article 36 [20], stated that increasing autonomy in weapons systems risks expanding the notion of an attack in ways that undermine the exercise of MHC. In the context of weapons that can detect and engage target objects, less control can be exercised by the weapon over the effects of the weapon if it (i) operates for a long time over a wider area, (ii) uses wider proxy indicators, and (iii) is used in an environment where there are a number of persons and objects that match those parameters.

This raises a number of important questions to which the CCW needs urgently to apply itself. Recognizing a requirement of MHC over individual attacks is likely to lead towards the prohibition of certain weapons systems, or certain uses of weapons. If we hold that arguing in favour of weapons that operate without MHC is morally and legally untenable, it is relatively straightforward to conclude that autonomous weapons, that do not allow MHC, should be banned. The CCW should establish key principles with reference to which the issue of MHC should be approached and then draw the boundaries of the necessary prohibitions. No delegation has argued that autonomous weapons should be allowed to operate without human control, or with human control that is devoid of meaning. Since the CCW is specifically designed to produce new prohibitions and restrictions on weapons, such initiatives should not be considered premature.

Lin (2015) was of the view that the Martens Clause requires that armed conflicts proceed under "the laws of humanity and the requirements of the public conscience." [21] The right to life, the right to human dignity, and the Martens Clause are all related. The killing of an individual using LAWS would violate a basic right to life—that is, a right not to be killed arbitrarily, unaccountably, or otherwise inhumanely. The Martens Clause expands our understanding of human dignity. It is fundamental to the thinking that has led to the prohibition of weapons that are *mala in se* or evil in themselves, such as poison, exploding bullets, blinding lasers and nuclear weapons in war. A soldier being chased and hunted by a robot would not only be inhumane but would also be violative of human dignity.

The ICRC was of the opinion that there has been a broad agreement on the need to retain human control over the critical functions of LAWS. It held that incrementally increasing autonomy in weapons systems – specifically in the critical functions of identifying, selecting, and attacking targets – has raised fundamental questions about human control over the use of force. Future discussions at the CCW, in view of the ICRC, should address whether these developments may affect the ability of parties to armed conflicts to respect IHL, and whether these developments are acceptable under the principles of humanity and the dictates of public conscience.

The State Parties to the 2015 CCW meeting decided to convene an informal meeting of experts in April 2016 to discuss the questions related to emerging technologies in the area of LAWS, in the context of the objectives and purposes of the Convention.

The 2016 CCW Informal Meeting of Experts

In the CCW meetings of experts from 11 to 15 April 2016, the states as well as NGOs[22] raised the issue of 'definition', 'responsibility' and 'accountability' with respect to LAWS. The panel on "towards a working definition" focused on different ways to define LAWS by their technical features. A number of states proposed that human control must be maintained over weapon systems, regardless of whether this should be considered as appropriate, meaningful or effective. It was of common understanding that, as with all weapon systems, the rules of IHL are fully applicable to LAWS.

There was a widely shared understanding that the responsibility for the development, production and deployment of LAWS rests with the operating state. However, some delegations were of the view that individuals could be held responsible under the relevant bodies of international law. The importance of ensuring an unequivocal accountability chain in the deployment of a weapon system was underlined. Some delegations raised serious security concerns, including the possibility of an arms race, potential for lowering of the threshold for the use of force, and exacerbation of global and regional instability. A number of delegations proposed a preventive approach, calling for a prohibition on the development, acquisition, trade, deployment and use of LAWS. Some also called for a moratorium for ongoing development and production processes.

The ICRC has recently stated that it will be a daunting task to ensure that future LAWS would be capable of distinguishing military objectives from civilian objects, combatants from civilians, and active combatants from persons *hors de combat*. The assessment of whether an autonomous weapon system can be used in compliance with IHL may depend on the specific technical characteristics and performance of the weapon system and the intended and expected circumstances of its use. According to the ICRC, certain technical characteristics and their interaction with different operational parameters could significantly affect this assessment. These are (i) the task the weapon system carries out; (ii) the type of target the weapon system attacks; (iii) the environment in which the weapon system operates; (iv) the movement of weapon system in space; (v) the timeframe of operation of the weapon system; (vi) the adaptability of the weapon system, i.e. its ability to adapt its behaviour to changes in its environment, to determine its own functions and to set its own goals; (vii) degree of reliability of the weapon system, i.e. robustness to failures and vulnerability to malfunction or hacking; and (viii) potential for human supervision and intervention to deactivate the weapon system. The combination of these technical characteristics and performance of the weapon system with the operational parameters of its use are critical to determining the foreseeable effects of the weapon – in other words, the predictability of the outcomes of using the weapon – and therefore in determining whether it can be used in conformity with IHL rules.

Under the law of state responsibility, in addition to accountability for violations of IHL committed by its armed forces, a State could be held liable for violations of IHL caused by an autonomous weapon system that it has not, or has inadequately, tested or reviewed prior to deployment. Under the laws of product liability, manufacturers and programmers could also be held accountable for errors in programming or for the malfunctioning of an autonomous weapon system. Though new technologies of warfare are not specifically regulated by IHL treaties, their development and employment in armed conflict does not occur in a legal vacuum. As with all weapon systems, LAWS must be capable of being used in compliance with IHL. The responsibility for ensuring this rests, first and foremost, with each State that is developing these new technologies of warfare.[23]

Conclusion

The recent developments in robotics have raised the possibility of removing human control over weapon systems. International concern over this possibility resulted in the meeting of experts to discuss LAWS in the framework of the CCW in April 2016. The experts articulated diverging views on LAWS. Some supported the idea of MHC over LAWS, while a few were of the opinion that the term was too vague. However, respect for international law, in particular of IHL and international human rights law, was universally appreciated at the meeting. The important elements of IHL referred to were an unequivocal accountability chain in the deployment of a weapons system, and respect for the principles of distinction, proportionality and precaution in attack. A number of delegates expressed the view that it was too early to draw far-reaching conclusions as the subject needed further clarification.

Endnotes

1   The Convention on Prohibition or Restriction on the Use of Certain Conventional Weapons Which May be Deemed to Be Excessively Injurious or to Have Indiscriminate Effects as amended on 21 December 2001 is usually referred to as the Convention on Certain Conventional Weapons (CCW). It has 121 State Parties and 5 signatories.

2   At the 2013 CCW Meeting of High Contracting Parties, a new mandate on lethal autonomous weapons systems (LAWS) was agreed on. The mandate stated "…Chairperson will convene in 2014 a four-day informal Meeting of Experts, from 13 to 16 May 2014, to discuss the questions related to emerging technologies in the area of lethal autonomous weapons systems, in the context of the objectives and purposes of the Convention. He will, under his own responsibility, submit a report to the 2014 Meeting of the High Contracting Parties to the Convention, objectively reflecting the discussions held." The CCW seeks to ban or restrict the use of specific types of weapons that are considered to cause unnecessary or unjustifiable suffering to combatants or to affect civilians indiscriminately.

3   For more details see compilations made by Campaign to Stop Killer Robots, Country Statements/ Policy Positions dated March 2014 and 23 March 2015.

4   See: Christof Heyns, Report of the Special Rapporteur on extrajudicial, summary or arbitrary executions, UNGA Doc A/HRC/23/47 dated 9 April 2013.

5   In May 2014, representatives from 87 States participated in the first CCW informal meeting of experts to consider questions related to emerging technologies in the area of lethal autonomous weapons systems.

6   Chemical, biological, radiological and nuclear defence (CBRN defence) is protective measures taken in situations in which chemical, biological, radiological or nuclear warfare (including terrorism) hazards may be present.

7   Switzerland is providing financial support to a research project by UNIDIR on autonomy in weapons systems.

8   The US Department of Defense Directive 3000.09 of November 2012

9   The ICRC has shown its concerns with the LAWS in October 2011 in a report entitled International Humanitarian Law and challenges to contemporary armed conflicts (presented to the 31st International Conference of the Red Cross and Red Crescent). It has also published a short "frequently asked questions" outlining its position and concerns with the LAWS. See: *International Humanitarian Law and the challenges of contemporary armed conflicts*, Report for the 31st International Conference of the Red Cross and Red Crescent, Geneva, 28 November to 1 December 2011, p. 39-40.

10   From 26 to 28 March 2014, the ICRC convened an international expert meeting entitled at Geneva. It brought together government experts from 21 States and 13 individual experts with a wide range of legal, technical, operational, and ethical expertise. The aim was to gain a better understanding of the issues raised by autonomous weapon systems and to share perspectives. See: *Autonomous weapon systems: Technical, military, legal and humanitarian aspects*, Expert meeting, Geneva, Switzerland, 26-28 March 2014. ICRC, Geneva.

11   The discussion falls under the purview of the CCW, which has five protocols already covering non-detectable fragments, mines and booby traps, incendiary weapons, blinding lasers and the explosive remnants of war. The 2014 Meeting of the High Contracting Parties to the Convention held on 13 and 14 November 2014 in Geneva decided, as contained in paragraph 36 of its final report (CCW/MSP/2014/9), "to convene under the overall responsibility of the Chairperson an informal meeting of experts of up to five days during the week of 13 to 17 April 2015 to discuss the questions related to emerging technologies in the area of lethal autonomous weapons systems, in the context of the objectives and purposes of the Convention.

12   The following High Contracting Parties to the CCW participated in the work of the meeting: Albania, Argentina, Australia, Austria, Belarus, Bolivia, Bosnia and Herzegovina, Brazil, Bulgaria, Canada, Chile, China, Colombia, Croatia, Cuba, Cyprus, Czech Republic, Ecuador, El Salvador, Estonia, Finland, France, Georgia, Germany, Greece, Guatemala, Holy See, Honduras, Hungary, India, Iraq, Ireland, Israel, Italy, Japan, Jordan, Kazakhstan, Kuwait, Lao People's Democratic Republic, Latvia, Lithuania, Madagascar, Mexico, Mongolia, Morocco, Netherlands, New Zealand, Norway, Pakistan, Philippines, Poland, Portugal, Qatar, Republic of Korea, Republic of Moldova, Russian Federation, Saudi Arabia, Serbia, Sierra Leone, Slovakia, Slovenia, South Africa, Spain, Sri Lanka, State of Palestine, Sweden, Switzerland, Tunisia, Turkey, Uganda, Ukraine, United Arab Emirates, United Kingdom of Great Britain and Northern Ireland, United States of America, Venezuela (Bolivarian Republic of) and Zambia. The observers included: Algeria, Brunei, Darussalam, Cote d'Ivoire, Ghana, Indonesia, Lebanon, Libya, Malaysia, Mozambique, Myanmar, Singapore, Thailand and Yemen. Egypt, the Signatory State to the CCW also participated in the work of the meeting.

13   The representatives of the following non-governmental organizations participated in the work of the meeting: Basel Peace Office, Campaign to Stop Killer Robots [Amnesty International, Article 36, Facing Finance, Human Rights Watch, SEHLAC, International Committee for Robot Arms Control (ICRAC), Mines Action Canada, Nobel Women's Initiative, Norwegian Peace Association, PAX, Pax Christi Ireland, Protection, Women's International League for Peace and Freedom (WILPF)], Centre for a New American Security, International Campaign to Ban Landmines–Cluster Munition

Coalition (ICBL–CMC), and World Council of Churches.

14 The representatives of the following entities also participated in the work of the meeting: Brigham Young University Law School; Centre for Land Warfare Studies; Columbia Law School; European University Institute; Geneva Academy of International Humanitarian Law and Human Rights (ADH); Geneva Centre for Security Policy; Graduate Institute of International and Development Studies; Josef Korbel School of International Studies, University of Denver; National University of Ireland; Peace Research Institute Frankfurt (PRIF); Stockholm International Peace Research Institute (SIPRI); The Harvard Sussex Program (SPRU) University of Sussex; University of Geneva; University of California; University of Central Lancashire; University of Strasbourg and University of Valenciennes.

15 Available at: http://www.unog.ch/80256EE600585943/(httpPages)/6CE049B E22EC75A2C1257C8D00513E26?OpenDocument, accessed 25 September 2015.

16 International Committee for Robot Arms Control (ICRAC) is an international not---for---profit association of scientists, technologists, lawyers, and policy experts Committed to the peaceful use of robotics in the service of humanity and the regulation of robot weapons. ICRAC is a founding member of the Campaign to Stop Killer Robot. For more information visit: <http://www. icrac.net.> and < http://www.stopkillerrobots.org.>.

17 Over 30 nations already operate human-supervised autonomous weapon systems to defend bases or vehicles against attacks from mortars, rockets, or missiles. Systems in this category include automated air and missile defense systems as well as active protection systems for ground vehicles. These systems have been used narrowly to defend human-occupied vehicles and bases and retain a person "on the loop" who supervises operation and can intervene, if necessary. Scharre Paul, Horowitz Machael C and Kalley Sayler, Autonomous Weapons at the UN: A Primer for delegates, Centre for New American Security, April 2015.

18 Scharre Paul, Horowitz Machael C and Kalley Sayler, Autonomous Weapons at the UN: A Primer for delegates, the Centre for New American Society (CNAS), April 2015. CNAS is an independent non-profit research institution.

19 Sayler Kelley, CNAS Statement to the UN Convention on Certain Conventional Weapons, 14 April 2015.

20 Article 36 is a UK-based NGO working to prevent the unintended, unnecessary or unacceptable harm caused by certain weapons. It undertakes research, policy and advocacy and promotes civil society partnerships to respond to harm caused by existing weapons and to build a stronger framework to prevent harm as weapons are used or developed in the future. The name refers to Article 36 of the 1977 Additional Protocol I that requires states to review

new weapons, means and methods of warfare. For more information, see: *www.article36.org.*

21 Lin Patrick, 'The right to life and the Martens Clause', presentation made at the Convention on Certain Conventional Weapons meeting of experts on LAWS, at United Nations in Geneva, Switzerland on 13-17 April 2015.

22 The following High Contracting Parties to the CCW participated in the work of the meeting: Albania, Algeria, Argentina, Australia, Austria, Belarus, Belgium, Bosnia and Herzegovina, Brazil, Bulgaria, Cambodia, Cameroon, Canada, Chile, China, Colombia, Croatia, Costa Rica, Cuba, Cyprus, Czech Republic, Djibouti, Dominican Republic, Ecuador, El Salvador, Estonia, Finland, France, Georgia, Germany, Greece, Holy See, Honduras, Hungary, India, Iraq, Ireland, Israel, Italy, Japan, Jordan, Kazakhstan, Kuwait, Lao People's Democratic Republic, Latvia, Lithuania, Mexico, Mongolia, Montenegro, Morocco, Netherlands, New Zealand, Norway, Pakistan, Panama, Peru, Philippines, Poland, Portugal, Qatar, Republic of Korea, Romania, Russian Federation, Saudi Arabia, Serbia, Sierra Leone, Slovakia, Slovenia, South Africa, Spain, Sri Lanka, Sweden, Switzerland, Tunisia, Turkey, Uganda, Ukraine, Uruguay, United Kingdom of Great Britain and Northern Ireland, USA, Venezuela, and Zambia. The following States not parties to the CCW participated as observers: Ghana, Indonesia, (Islamic Republic of Iran), Lebanon, Malaysia, Myanmar, Oman, Singapore, Thailand, Yemen and Zimbabwe. The representatives of the United Nations Institute for Disarmament Research (UNIDIR), United Nations Interregional Crime and Justice Research Institute (UNICRI), United Nations Office for Disarmament Affairs (UNODA), European Union, International Committee of the Red Cross (ICRC), International Federation of Red Cross and Red Crescent Societies (IFRC) and Geneva International Centre for Humanitarian Demining (GICHD) participated in the work of the meeting. The representatives of the NGOs participated in the work of the meeting were: Campaign to Stop Killer Robots, Amnesty International, Article 36, Association for Aid and Relief, Japan, Facing Finance, Human Rights Watch, International Committee for Robot Arms Control (ICRAC), Mines Action Canada, Nobel Women's Initiative, Nonviolence International, Norwegian Peace Association, PAX, Project Ploughshares (Canada), Pugwash Conferences on Science and World Affairs, SEHLAC, Women's International league for Peace and Freedom, Centre for a New American Security, Human Rights Now, Human Rights Watch, and International Campaign to Ban Land mines - Cluster Munitions Coalition (ICBL-CMC).

23 Report: International humanitarian law and the challenges of contemporary armed conflicts, 32IC/15/11, Document prepared by the International Committee of the Red Cross, Geneva, October 2015.

VIII Conclusion and the Way Ahead

Rapid development in robotic technology has led to the introduction of unique autonomous weapons systems in the last decade. The level of autonomy in these weapons has been growing steadily and there are several weapons systems approaching fully autonomous capabilities. Today, South Korea's SGR-A1 sentry robots guard the demilitarized zone with human oversight. Israel has Sentry Tech systems along the Gaza border which have an autonomous firing mode. Israel also has the Guardium, which patrols the international border and can respond to stimuli in the environment. The US Navy's PHALANX system is able to autonomously detect, track and fire at anti-ship missiles. The US Navy's X-47B has been developed to take off, land and refuel on an aircraft carrier without human oversight. The US military SWARM technology, when fully developed, would allow one human operator to control a fleet of aircraft, which have been designed to respond in a synchronized fashion. A few other countries are not far behind in the development of such weapon systems. It is believed that the introduction of LAWS in the battlefield is inevitable and imminent. The former chief scientist of the US Air Force has stated that the technology required for fully autonomous military strikes already exists.

States are also investing significantly in increasing the operational autonomy of weapon systems so that their less lethal versions can be adopted by law enforcement agencies. These autonomous systems may be used to launch tear gas, rubber bullets or paint markers, deliver electric shocks, or discharge firearms. Equipped with cameras, they could also be used to apprehend escaping prisoners, secure prisons, ensure perimeter security, and identify rioters or those transgressing the law.[1] Spain, for instance, has successfully developed a robotic weapon called the "RiotBot", the first robot in the world to be used specifically for riot control.[2] With a speed of over 18 kmph and a weight of less than 20 kg, it can be deployed and used quickly and safely by one operator. The RiotBot employs a NLS 900 carbine, modified and adapted for safe use in the robot. It has a shooting velocity of 900 balls per minute and a total capacity of 450 PAVA balls, a non-lethal ammunition.[3]

The proponents of LAWS argue that, once fully developed, LAWS would be deployed in combat situations. For instance, initially they may be deployed in exclusive battlefields for attacking combatants and destroying military objectives. They may also be used in naval warfare, in which the possibility of collateral damage would be remote. LAWS could also operate underwater to patrol for and attack enemy submarines without the risk of causing excessive collateral damage.[4] They could be used in the air to patrol a no-fly zone, hunting for enemy aircraft where no civilian aircraft are permitted to fly. The weapon may be programmed to recognize the profiles of enemy aircraft that would distinguish them from civilian aircraft. When used in such situations, LAWS would be able to adhere to the principles of distinction[5] and proportionality[6] as no civilians or civilian objects would be present in the combat zone. Further, as the technology progresses, it would be possible for LAWS to comply with the principles of distinction and proportionality in other combat situations also.[7] Autonomous weapons would be much cheaper compared to the massive expense necessary for manufacturing and maintaining manned weapon systems. Besides, their use would pose no risk to armed forces personnel and the number of war casualties is likely to be very small.

On the other hand, the opponents of LAWS firmly believe that such weapons would never have the necessary skills to discriminate between combatants and innocents. Such weapons, if developed, would not be able to follow two basic principles of IHL, i.e., distinction and proportionality. Distinction entails that persons employing force must always distinguish between lawful military targets, like enemy combatants, equipment or facilities, and protected persons, like civilians, medical personnel or persons who are *hors de combat,* and civilian property. Proportionality prohibits combatants from launching attacks against military targets if the collateral damage is 'excessive' compared to the military advantage that will be gained by the attack. Allowing such weapons to make decisions about who to kill would fall foul of the fundamental ethical precepts of a just war under *jus in bello* as enshrined in the Geneva and Hague Conventions and the various protocols drawn up to protect civilians, wounded soldiers, the sick, the mentally ill and captives.

According to Krishnan, there are three main concerns regarding the ability of LAWS to distinguish legal targets from civilian targets: (i) they may be susceptible to weak machine perception; (ii) they may have difficulty interacting with their environment, leading to the frame problem; and (iii) there may be a problem of weak software.[8] The major drawback of autonomous systems is that with the increase in software complexity, there will be more opportunities for bugs and vulnerabilities; and as these systems are injected into an adversarial environment, there will be opportunities for encountering situations that the original designers had never considered.

Theoretically, LAWS may have the potential to apply force more proportionately and appropriately than a combatant.[9] However, the ability of an autonomous weapon system to apply force proportionately will ultimately depend on the weapons with which it is equipped. Since proportionality is concerned with preventing excessive collateral damage, human judgement relating to situation awareness, adaptability and the capability to react swiftly to changing circumstances will always be a crucial factor in taking such decisions.

The possibility of malfunctioning of LAWS cannot be ruled out. These weapon systems would also be susceptible to spoofing, thus allowing unauthorized actors to control them for illegitimate purposes. The enemy might be able to use cyber means to take control of LAWS and direct it against friendly forces or a civilian population. Schmitt, though, rules out such possibilities. Referring to the US policy of developing such systems, he states, "Those developing the systems are acutely aware of such risk."[10] However, a few cases of hacking of lethal drones[11] shows that even LAWS would always be susceptible to such system weaknesses.

It would also be difficult for LAWS to identify whether an enemy solider has become *hors de combat* or if the situation is merely deceitful. An example is if a lethal autonomous weapon identifies an enemy combatant and attacks, but the attack is unsuccessful; however, the combatant fakes an injury. Would it be possible for LAWS to correctly identify such an enemy as *hors de combat*? Such assessment in the fog of war is difficult even for a human soldier; nevertheless, a human solider will be able to assess the entire context, whereas LAWS, because of programming limitations, may not have the capability to assess such situations.[12]

Today's robotic machines are capable of only limited reasoning due to the fact that their computations are specified. Everything a robot does is spelt out with simple instructions, and the scope of its reasoning is entirely contained in its programme. A robot's perception of its environment through its sensors is quite limited. If it encounters circumstances that it has not been programmed to handle or that fall outside the scope of its capabilities, it may behave abnormally.[13] Another concern with LAWS is that the robot creators tend to be engineers, programmers and designers with little training in ethics, human rights, IHL or security. They may be even unaware of what ROE means. According to Nourbakhsh, Professor of Robotics at the Robotics Institute of Carnegie Mellon University, in the US, hardly any of the academic engineering programmes that grant a degree in robotics require an in-depth study of such fields.[14]

Today, the artificial intelligence and robotics communities face an important ethical decision: whether to support or oppose the development of LAWS. It has been described as the third revolution in warfare, after gunpowder and nuclear weapons, and could be a reality in the next two decades. The law of war or IHL, which governs the conduct of hostilities, has no specific provisions for LAWS. The rules of IHL require that an attack must satisfy three criteria: military necessity; discrimination between combatants and non-combatants; and proportionality between the value of the military objective and the potential for collateral damage.

Concerned by the likely development of LAWS and their deployment in the battlefield, certain non-governmental organizations, like the Campaign to Stop Killer Robots, Human Rights Watch, Article 36, International Committee for Robot Arms Control (ICRAC), and Nobel Women's Initiative, have led a campaign for a moratorium on the development of such weapon systems. The ICRAC came out with a declaration calling for a ban on killer robots. This declaration has been signed by over 270 roboticists, scientists and other experts.

The CCW Mechanism

In 2013, the State Parties to the UN Convention on Certain Conventional

Weapons (CCW) began a process to study LAWS as the first step towards a new protocol for controlling or prohibiting such weapons. In April 2013, UN Special Rapporteur Christof Heyns called for a moratorium on the development of LAWS, pending the formation of an experts' panel to formulate a policy. During meetings at the CCW in 2014, 2015 and 2016, several States expressed opinions regarding the development and use of LAWS in an armed conflict. Their views varied. A few States recommended a "total ban" on their development, while a few adopted a policy of "wait and watch" as these weapons were yet to be fully developed, and a prohibition would be premature, unnecessary or unenforceable. At the same time, a few States were of the opinion that the human rights aspects of the issue should be addressed as the CCW might not be the only appropriate framework for a discussion on LAWS. The States in the first category regarded LAWS as fundamentally in conflict with the basic principles of IHL, like proportionality and distinction; and called for an immediate, legally binding instrument providing for a ban. In their opinion, such a ban would encompass the development, acquisition and deployment of and trade in LAWS. It was argued that LAWS would be unethical by their very nature as they lack human judgement and compassion. LAWS could generate new risks of proliferation and lead to a new arms race. They could challenge regional balances and possibly global strategic stability as well as affect general progress on disarmament and non-proliferation. They could also lower the threshold for starting or escalating military activity. Finally, they could fall into the hands of non-state actors and increase the risk and potential of terrorism.[15] There are confirmed reports that the Islamic State (IS) has built a research centre devoted to launching attacks in the West using driverless cars and rehabilitated anti-aircraft missiles. [16]

The opinion expressed by the Indian delegation at the 2014 CCW expert meeting is worth mentioning: "India attaches importance to the CCW as a living and dynamic Convention....From India's point of view, we would like the CCW process to emerge strengthened from these discussions, resulting in increased systemic controls on international armed conflict embedded in international law in a manner that does not further widen the technology gap amongst states or encourage the use of lethal force to settle international disputes just because it affords the prospects of lesser causalities to one side or that its use can be insulated from the dictates of public conscience. Overall, the consideration of this issue is a test case of whether the CCW can respond meaningfully to evolving new technology as applicable to armed conflict in this century."[17]

According to the German Ambassador Michael Biontino, who chaired the April 2015 meeting, "Machines or systems tasked with making fully autonomous decisions on life and death without any human intervention, were they to be developed, would be in breach of IHL, unethical and possibly even pose a risk to humanity itself." Since there is no agreement on what constitutes "human intervention", this statement is of little value.[18] A few highlights of the discussion at the CCW experts meeting are as follows.

- Autonomy should be seen as a characteristic of technology attached to a weapon system, not the weapon itself. Therefore, LAWS is an "umbrella term" covering a wide range of weapon systems.

- Increasing the autonomy and complexity of weapons systems may lead to outcomes that are less predictable. Deploying a weapon system with unpredictable effects creates a significant risk of breach of IHL.

- At the operational stage, a commander would have to fully understand the capabilities of the autonomous system to make a judgement on the acceptable risk in deployment.

- A rigorous legal review of such weapons is needed. A significant challenge would be how to test LAWS as part of such a process and in particular, how to test predictability.

- The States must establish a mechanism for review of new weapons and methods of warfare.

- The use of LAWS would potentially affect the right to life, the right to bodily integrity, the right to human dignity, the right to humane treatment and the right to remedy.

- Even if a machine were able to comply with the requirements of IHL, decisions taken by machines over life and death would affect the dignity of the person targeted. Even if LAWS were more accurate than soldiers and thus spared more lives in comparison to soldiers, the dignity of those targeted would still be affected.

- Autonomous technologies could lead to more discriminating weapons systems. It may therefore be premature to prohibit

LAWS on the basis of the current shortcomings in autonomous technologies.

The delegates who claimed that the prohibition of LAWS without a clear understanding of the potential opportunities and risks of the technology was premature were perhaps not correct in their approach.[19] The problems posed by future LAWS must be discussed today, while the technology is still at a relatively nascent stage and there is an opportunity to control its development. Research and development in the field of autonomous weapons has reached a critical stage, making it imperative to have an in-depth analysis on the need for the development of such weapon systems. We cannot wait to see the effectiveness of such weapons and then try to ban them.

LAWS in Non-international Armed Conflict

The future use of LAWS in non-international armed conflict (NIAC) cannot be ruled out. Drones have already been used in counterterrorism operations by certain States, so it is not unthinkable that LAWS may also be deployed in NIAC when they become available. In Pakistan, the government had initially collaborated with the US military and allowed them to use armed drones in Pakistan territory against terrorists.[20] With the elimination of several militants, the minimizing of US military casualties, and decreasing public concern about military action in Pakistan, armed drones offered the US a win-win tactical solution in the war on terror.[21] However, hundreds of innocent civilians and children were also killed in the drone attacks and the civilian population suffered great mental trauma. Military analysts are of the view that governments would prefer to deploy LAWS on a permanent basis in NIAC zones, to free their paramilitary and military forces for additional tasks. They also believe that the deployment of LAWS would eliminate risk to their military personnel in NIAC.

The likely use of LAWS in NIAC raises serious human rights concerns. It would threaten the right to life, the prohibition of torture and other cruel, inhuman or degrading treatment or punishment, the right to security of person and other human rights. It would also violate Principle 9 of the 1990 Basic Principles on the Use of Force and Firearms by Law Enforcement Officials, which mandates that intentional lethal force is only lawful when strictly unavoidable to protect life.[22] Robotization will also run contrary to the current doctrine of NIAC, wherein governments are following the policy of winning over hearts and mind (WHAM).[23] The existing principles of customary international law forbid the use of LAWS in all contexts.[24]

Human Rights Issues

Article 36 of AP I provides, "In the study, development, acquisition or adoption of a new weapon, means or method of warfare, a High Contracting Party is under an obligation to determine whether its employment would, in some or all circumstances, be prohibited by this Protocol or by any other rule of international law applicable to the High Contracting Party." The words "or by any other rule of international law" may be interpreted to imply that in order to pass an Article 36 review, the potential use of a weapon must also be considered with reference to the applicable human rights law, including the right to dignity. The ICRC commentary on AP I provides that on the basis of Article 36, the State Parties should undertake to determine the possibly unlawful nature of a new weapon, both with regard to the provisions of the Protocol, and with regard to any other applicable rule of international law.[25] The determination is to be made on the basis of the normal use of the weapon as anticipated at the time of evaluation. If these measures are not taken, the State will be responsible for any wrongful damage ensuing.[26]

Within the military, paramilitary or police establishments, there are many layers of delegated authority, from the highest down to the soldier or constable, but at every level there is a responsible human being to bear both the authority and responsibility for the use of force. The nature of command responsibility does not allow one to abdicate one's moral and legal obligations to determine that the use of force is appropriate in a given situation. One might transfer this obligation to another responsible human agent, but one then has a duty to oversee the conduct of that subordinate agent. Autonomous lethal systems are not responsible human agents and one cannot delegate this authority to them in an armed conflict or a law enforcement situation.

It may be claimed by the proponents of LAWS that the use of LAWS would reduce the stress and trauma of combatants by keeping them away from the war zones. While the use of LAWS may reduce physical trauma and some mental and emotional trauma, it is possible that it would generate significant psychological stress related to the operator being simultaneously "present in and absent from" the battlefield. Through their link with LAWS, combatants may witness events that are psychologically distressing. They may find themselves witnessing events in which they are powerless to intervene. For instance, a combatant may witness the massacre of civilians, or other war crimes being committed by LAWS, yet be helpless to prevent them. According to Sparrow, there is a strong possibility that the operation of LAWS would expose the operators to unique psychological stresses.[27]

The International Committee of the Red Cross

The Common Article 1 to the four Geneva Conventions of 1949 lays down an obligation to respect and ensure respect for the Conventions in all circumstances. Every State signatory to the Geneva Conventions, therefore, not only has an international legal obligation to ensure strict compliance with the provisions of the Conventions, but also a duty to take measures to ensure that the means and methods adopted by its armed forces do not violate the basic principles of IHL.[28]

The ICRC, which has often voiced its opinion on issues related to the development and the use of new weapons, has surprisingly said very little about LAWS. In September 2011 the President of the ICRC acknowledged the need to focus on robotic weapons systems in the future. However, he was of the view that such systems might be capable of enhancing the level of protection available. In his view, the deployment of LAWS might cause fewer incidental civilian casualties and less incidental civilian damage compared to the use of conventional weapons.[29]

In its subsequent report, the ICRC acknowledged that the level of autonomy in weapon systems is expected to increase to reduce the workload of operators and increase the time available for them to focus on decision-making. According to the ICRC, it was not clear whether such technologies are likely to be sufficiently sophisticated to enable autonomous functioning in highly complex decision-making, such as the process of selecting and attacking targets. It held that in a discussion of LAWS, it may be useful to focus on autonomy in critical functions, rather than autonomy in the overall weapon system. According to the ICRC, the term 'autonomous weapon systems' refers to weapon systems for which critical functions (i.e. acquiring, tracking, selecting and attacking targets) are autonomous and ensuring accountability for acts of LAWS poses some significant challenges.[30]

The ICRC, in its recent report has become more vocal on the issue of LAWS. It is of the view that although new technologies of warfare are not specifically regulated by IHL treaties, their development and employment in armed conflict does not occur in a legal vacuum. Thus, LAWS must be capable of being used in compliance with IHL, in particular its rules on the conduct of hostilities. The responsibility for ensuring this rests, first and foremost, with each State that is developing these new technologies of warfare. Further, new weapons should not be employed prematurely under conditions in which respect for IHL cannot be guaranteed.[31] While, it may not be possible for the ICRC to stop the development of a new weapon system by a State, it can try to impress upon the States to ensure that every such weapon goes through the Article 36 review and is used in compliance with IHL.[32]

According to the ICRC, there are three broad approaches that states could take to address the legal and ethical questions raised by LAWS: (i) strengthening national mechanisms for legal review and implementation of IHL to ensure any new weapons, including LAWS, can be used in compliance with IHL; (ii) developing a definition of LAWS in terms of the weapon systems that may be problematic from a legal and/or ethical perspective with a view to establishing specific limits on autonomy in weapon systems; and (iii) developing the parameters of human control in light of the specific requirements under IHL, thereby establishing specific limits on autonomy in weapon systems.

According to O'Connell, the possession of technology lowers the political and psychological barriers to killing, making it easier to overlook the legal, policy and ethical limits as well.[33] Allowing the development and the use of LAWS creates a moral and ethical dilemma similar to that of the nuclear weapons. Just as the use of nuclear power can provide clean, efficient energy, robotics has limitless potential to help humanity in various spheres. On the other hand, LAWS, like nuclear weapons, could cause indiscriminate harm to civilians. Even if used only against combatants, it would be unethical to allow a machine to target and kill a human being. Deploying LAWS would also cause collateral damage and environmental harm, as we have already witnessed in the case of armed drones.

The Way Ahead

In response to activism surrounding the legal and ethical dimensions of LAWS that may control their own movement, detect their own targets, and make their own decision to fire at a target and kill, without human intervention, the CCW States Parties held discussions in April this year.[34] During the earlier negotiations, the States with access to advanced technology have argued that LAWS being developed by them would be able to meet the ethical challenges as well as abide by the rules of IHL and they should, therefore, be exempted from the ban. Such arguments must be rejected because the exemption would be based on the grounds of the availability of sophisticated and expensive technology that is not available universally. During the Oslo Diplomatic Conference that negotiated the Ottawa Antipersonnel Landmines Treaty, the US had pushed for a similar exemption for landmines equipped with self-destruction and self-deactivation features.

Some developed States may also thwart negations to ban LAWS on the ground that the proposed treaty poses dangers to their sovereignty and would reduce the effectiveness of their armed forces. The US has refused to be a signatory to the 2008 Convention on Cluster Munitions (CCM) maintaining that cluster munitions are legitimate weapons with clear military utility, and its "national security interests cannot be fully ensured consistent with the terms" of the CCM.[35]

Since almost all States Parties to the CCW wish to ensure that there is meaningful human control over the use of lethal force in targeting and engagement decisions made by weapons, they must make it explicit that human control is required over every individual attack. They must remember that only human beings are bound by the law. Angela Kane, the UN Representative for Disarmament Affairs, has voiced her opinion against the development of LAWS, stating, "there cannot be a weapon that can be fired without human intervention" because such weapons would bring forth "a faceless war....it should be outlawed."[36] A few organizations involved in artificial intelligence research and design have pledged not to accept any work related with the development of LAWS.[37]

The legal review of new weapons is essentially a national process, which may not be followed by all the States developing or acquiring them. There are also doubts whether all States have the necessary technical and scientific capabilities to effectively implement a review process. Since Article 36 does not provide standard methodology for the review, the States are likely to follow different standards during the weapon review processes. In addition, the States developing new weapons may not test such weapon systems thoroughly, as their results are not publicly disclosed. The States developing new weapon systems should, therefore, share their legal weapons review procedures with other CCW States Parties as a confidence-building measure.

The members of civil society could also play a significant role in the analysis and banning of future LAWS. In most states, the decision to research and develop a weapon system remains within the confines of government departments, and even the military may not be consulted in the initial stages of its development. If the issue is raised for discussions at an international forum, a select group of government officials take part in it. This coterie of officials jealously blocks the flow of information from the civil society to the government.[38] Members of the civil society have the duty to raise concerns about weapon technologies that are likely to have a wider negative impact on human health, environment and infrastructure. A case in point is the use of depleted uranium (DU) weapons by the US military in Gulf Wars and in Iraq and Afghanistan.[39]

Only time can tell whether future lethal robots, would ever be able meet the aspirations of humanity. For the moment, it seems unlikely that an autonomous weapon system can be morally superior to a human soldier on the basis of its being technologically capable of making fewer errors in a discrimination task, or minimizing disproportionate harms. An automated process must be subject to human review before it can legitimately initiate the use of lethal force. LAWS, no matter how good, cannot completely replace the presence of a true moral agent in the form of a human being possessed of a conscience and the faculty of moral judgment.

While one may argue that banning LAWS before it is actually used in the battlefield is illogical, it would be useful to recall that blinding laser weapons were banned before their development.[40] We have to ensure that new weapon systems comply with the fundamental moral principles underlying IHL and human rights law and their users remain accountable in case of any harm to protected persons. The European Parliament has already called for a moratorium or ban on LAWS.[41] In the unfaltering interest of humanity, international NGOs, the States, and the United

Nations must come together and create a treaty to place a comprehensive ban on the development and use of LAWS. A draft protocol to the CCW--Protocol VI--to ban the development and the use of LAWS has been placed at Appendix B to this book.

Endnotes

1   C. Heyns, 'Autonomous weapons systems and human rights law', Presentation to the informal CCW expert meetings, Geneva, 13–16 May 2014; Weizmann Nathalie, *Autonomous Weapon Systems under International Law,* Geneva Academy of International Humanitarian Law and Human Rights, November 2014, p. 27.

2   Information available at: http://www.technorobot.eu/en/riotbot_specifications.htm, accessed 15 October 2015.

3   The RiotBot can continuously operate alone for over two hours. The compressed air recharging, the replacement of batteries and the reloading of ammunition can all be done in less than five minutes. The remote control that controls the robot's movement is totally ergonomic and light-weight. The aim of the RiotBot is double—on the one hand, it does away with the risks incurred by the operative teams and on the other, it minimizes the impact of the intervention as it is a non-lethal weapon. For more details, see: http://www.technorobot.eu/en/riotbot_specifications.htm.

4   Groves *Steven,* The US Should Oppose the UN's Attempt to Ban Autonomous Weapons, Backgrounder, No. 2996, 5 March 2015.

5   There are general principles in IHL which ban indiscriminate and/or disproportionate attacks and there are treaties which ban the use of certain types of weapons because these weapons are considered either to be indiscriminate or cause disproportionate harm. Dum-dum bullets were banned because they caused disproportionate harm to combatants. Anti-personnel land mines and cluster munitions are banned because their effect is indiscriminate. Some of the protocols in the 1980 Convention on Certain Conventional Weapons (CCW) ban the use of weapons, like those which disperse undetectable fragments and blinding laser weapons. These are considered to cause disproportionate injuries. Barring nuclear weapons, international law bans indiscriminate and/or disproportionate attacks no matter what weapon is used.

6   The second IHL issue with LAWS is that they do not have the situational awareness or agency to make proportionality decisions. The principle of proportionality dictates that even if a weapon meets the test of distinction,

any actual use of a weapon must also involve an evaluation that sets the anticipated military advantage to be gained against the anticipated civilian harm (to civilian persons or objects). The harm to civilians must not be excessive relative to the expected military advantage.

7   Groves *Steven*, The US Should Oppose the UN's Attempt to Ban Autonomous Weapons, Backgrounder, No. 2996, 5 March 2015.

8   Krishnan Armin. 2009. *Killer Robots: Legality and Ethicality of Autonomous Weapons*, Ashgate Publishing Limited, p. 98–99.

9   Thomas, Bradan T., Autonomous weapon Systems: The Anatomy of Autonomy and the Legality of Lethality, *Houston Journal of International Law*, Vol. 37, No. 1, 2015, p. 235–274.

10 Schmitt, Michael N., Autonomous Weapon Systems and International Humanitarian Law: A Reply to the Critics, *Harvard National Security Journal Features*, 2013, p. 7.

11 On 4 December 2011, the US RQ-170 Sentinel stealth drone was captured by Iranian forces near the city of Kashmar in Iran. The Iranian government announced that the drone was brought down by its cyber warfare unit, which commandeered the aircraft and landed it safely. The US government initially denied the claims but later acknowledged that the downed aircraft was a US drone and requested that Iran return it. Hartmann Kim and Christoph Steup, The Vulnerability of UAVs to Cyber Attacks-An Approach to the Risk Assessment, 2013, available at: https://ccdcoe.org/cycon/2013/proceedings/ d3r2s2_hartmann.pdf, accessed 14 November 2015.

12 Ghasemi Ajda Hosseini, Semi-Autonomous Weapon Systems in International Humanitarian Law, Unpublished Master's Thesis, Faculty of Law, University of Lund, 2014, p. 20.

13 Rus Daniela, The Robots Are Coming: How Technological Breakthrough Will Transform Everyday Life, *Foreign Affairs*, July/August 2015, p. 4–5.

14 Nournakhsh Illa Reza, The Coming Robot Dystopia, *Foreign Affairs*, July/ August 2015, p. 23.

15 Meeting of the High Contracting Parties to CCW, Report of the 2015 Informal Meeting of Experts on Lethal Autonomous Weapons Systems (LAWS), UN Doc. CCW/MSP/2015/3 dated 2 June 2015, para 19.

16 Chulov Martin, Inside the ISIS terrorism workshops: video shows Raqqa research centre, The Guardian, 6 January 2016, http://www.theguardian.com/ world/2016/jan/06/inside-isis-terrorism-workshops-video-shows-raqqa- research-centre, accessed 7 January 2016.

17 India has stressed the continued relevance of the CCW in addressing challenges posed by the development and the use of new weapons and their systems with respect to international law in particular IHL. See: Statement by Permanent Representative of India to the Conference on Disarmament, at the CCW Experts Meeting on LAWS, 13 May 2014; and by Ambassador D B Venkatesh Varma, Permanent Mision of India on 11 April 2016.

18 Grimstad Lene, The UN's Meetings on Autonomous Weapons: Biting the Bullet, or Lost in Abstraction? Available at: http://www.isn.ethz.ch/Digital-Library/Articles/Detail/?lng=en&id=191992, accessed 5 September 2015.

19 Report of the 2015 Informal Meeting of Experts on LAWS, Meeting of the High Contracting Parties to the Convention on Prohibition or Restrictions on the Use of Certain Conventional Weapons (CCW) Which May be Deemed to Be Excessively Injurious or to Have Indoctrinate Effectsne, Geneva. UN Doc. CCW/MSP/2015/3, 2 June 2015, para 57.

20 Though the present government in Pakistan had never consented to the use of drones by the US military, consent was accorded during Pervez Musharraf's regime. Musharraf has recently admitted to having given the CIA permission to launch drone attacks inside his country. Jon Boone and Peter Beaumont, 'Pervez Musharraf admits permitting' 'a few' US drone strikes in Pakistan', *The Guardian*, April 12, 2013, available at: http://www.guardian.co.uk/world/2013/apr/12/musharraf-admits-permitting-drone-strikes, accessed 13 April 2013. Also see Ahmed Ishtiaq. 2013. *The Pakistan Military in Politics: Origin, Evolution, Consequences*, New Delhi: Amaryllis, p. 393.

21 Deri Aliya Robin, "Costless" War: American and Pakistani Reactions to the US Drone War, *Intersect*, Vol. 5, 2012, p. 1-15.

22 Principle 9 of the 1990 UN Basic Principles for the Use of Force and Firearms by Law Enforcement Officials: Law enforcement officials shall not use firearms except in self-defence or defence of others against the imminent threat of death or serious injury,….In any event, intentional lethal use of firearms may only be made when strictly unavoidable in order to protect life.

23 In Sub-conventional operations…….It must always be remembered that populace constitutes the centre of gravity of such operations and, therefore, winning of their hearts and minds is central to all our efforts during conflict management and resolution. Foreword by the Chief of the Army Staff, Doctrine for Sub-conventional Operations, Integrated Headquarter of Ministry of Defence (Army), Headquarters Army Training Command, Shimla, December 2006.

24 Ekelhof Merel and Miriam Struyk, Deadly Decisions: Eight Objections to Killer Robots, PAX, The Netherlands, 2014, p. 12-15.

25 The purpose of Article 36 (AP I) is to prevent states from developing inhumane weapons. Although all states have not ratified AP I, the requirement that the legality of all new weapons, means and methods of warfare be systematically assessed is customary in nature. Only a few states have put in place mechanisms to conduct legal reviews of new weapons and the review process is not always transparent and authentic, particularly in the case of dual-use technologies. For instance, in the case of the MQ1-Predator, the US Judge Advocate General (JAG) had first passed the Predator for surveillance missions. After it was armed with Hellfire missiles, the JAG said that because it had already separately passed both the Predator and Hellfire missiles, their combination did not require a new review. Ekelhof Merel and Miriam Struyk, Deadly Decisions: Eight Objections to Killer Robots, PAX, The Netherlands, 2014, p. 20.

26 Sandoz Yves, Swinarski Christophe and Zimmermann Bruno, Commentary on the Additional Protocols of 8 June 1977 to the Geneva Conventions of 12 August 1949, Geneva: ICRC, 1987, p. 423-425.

27 Sparrow Robert, Building a Better WarBot: Ethical issues in the design of unmanned systems for military applications, Science and Engineering Ethics, *Science and Engineering Ethics*, Vol. 15 (2), 2009, p. 169–187.

28 Common Article 1 (CA 1) to the Geneva Conventions of 1949 provides, "The High Contracting Parties undertake to respect and to ensure respect for the present Convention in all circumstances." The State have endorsed this interpretation of CA 1 during the International Conference of the Red Cross and Red Crescent, where they emphasizes: "the obligation of all States to refrain from encouraging violations of IHL by any party to an armed conflict and to exert their influence, to the degree possible, to prevent and end violations, either individually or through multilateral mechanisms, in accordance with international law." 30th International Conference of the Red Cross and Red Crescent, Resolution 3, 2007, Geneva: ICRC, para 2.

29 International Humanitarian Law and the challenges of contemporary armed conflicts, Official working document of the 31st International Conference of the Red Cross and Red Crescent, 28 November-1 December 2011, Geneva: ICRC.

30 *Autonomous weapon systems: Technical, military, legal and humanitarian aspects*, Expert meeting, Geneva, Switzerland, 26-28 March 2014.

31 International humanitarian law and the challenges of contemporary armed conflicts, Report, 32nd International Conference of the Red Cross and Red Crescent, ICRC: Geneva, October 2015, 32IC/15/11.

32 According to the ICRC, the deployment of LAWS "would reflect a paradigm shift and a major qualitative change in the conduct of hostilities." The ICRC

has called on nations not to employ LAWS unless compliance with LOAC can be "guaranteed." Autonomous Weapon Systems–Q & A, Geneva: International Committee of the Red Cross, 12 November 2014, available at: https://www. icrc.org/en/document/autonomous-weapon-systems-challenge-human-control-over-use-force#.VHyaxDHF98E, accessed 3 June 2015.

33 O'Connell Mary Ellen, The Future of Peace, Weapons, and War, available at: http://sg.tudelft.nl/wp-content/uploads/2015/06/The-Future-of-Peace-Weapons-and-War-Mary-Ellen-OConnell.pdf, accessed 15 December 2015.

34 The informal meeting of experts on LAWS was held from 11 April to 15 April 2016 at the Palais des Nations, Geneva.

35 Cluster munitions consist of a container that releases a large number of submunitions. It can be delivered by several means, which results in a shower of small explosives covering a large area. Cluster munitions, like other unguided ordnance, have the potential to miss their targets, thus harming civilians and their properties. Cluster submunitions are infamous for their high dud rates, which may be as high as 30-35 percent. These unexploded submunitions, become *de facto* landmines that kill and maim indiscriminately long after the conflict has ended. To protect civilians from the effects of cluster munitions, the Oslo Process resulted in the adoption of the Convention on Cluster Munitions in 2008. It prohibits the use, production, transfer, and stockpiling of cluster munitions. The US remains a leader in the development, use and export of cluster munitions. Andrew Feickert Andrew and Kerr Paul K., Cluster Munitions: Background and Issues for Congress, Congressional Research Service, 29 April 2014.

36 Ben Farmer, Killer Robots a Small Step Away and Must Be Outlawed, Says Top UN Official, *The Telegraph*, 27 August 2014.

37 Getting to Know: Ryan Gariepy, Arms Control Today, September 2015, p. 40.

38 Rappert Brian, Moves Richard, Anna Crowe, Nash Thomas, The roles of civil society in the development of standards around new weapons and other technologies of warfare, *International Review of the Red Cross*, Vol. 94, No. 886, Summer 2012, p. 765-786.

39 In the Gulf Wars the US introduced armour-piercing ammunition made of depleted uranium (DU), a radioactive and toxic waste. By the end of the war, nearly 300,000 kg of DU contaminated the soil on the battlefields of Saudi Arabia, Kuwait, and southern Iraq. The use of DU munitions, poses various short- and long-term hazards to the health of local populations, as well as to the environment. Exposure to large quantities of DU oxide is likely to cause an increase in the incidents of cancers, a finding that has been supported by the World Health Organization. *Depleted Uranium: Source, Exposure and Health Effects*, 2001, Geneva: World Health Organization, p. 84.

40 Protocol IV of 1995 to the Convention on Prohibitions or Restrictions on the Use of Certain Weapons Which May Be Deemed to Be Excessively Injurious or to Have Indiscriminate Effects bans the development of lasers that were "specifically designed" to cause permanent blindness. It was based on the idea that blindness was "superfluous injury or unnecessary suffering" in the context of an armed conflict. The Protocol allowed the development of lasers for targeting in armed conflict if it did not involve blindness.

41 The European Parliament has adopted a resolution [2014/2567 (RSP)] for a ban on the development, production and use of LAWS which enable strikes to be carried out without human intervention. Ninety percent of the Members of the European Parliament (MEPs) who voted were in favour of the resolution, while just eight percent of MEPs voted against.

Appendix A

This appendix contains the provisions of IHL, which are relevant to the use of LAWS. These provisions have been drawn from the 1907 Hague Regulations; the 1949 Geneva Conventions Relative to the Protection of Civilians in the Time of War; the 1977 Additional Protocol I, relative to the Geneva Conventions of 1949; the 2003 Rome Statute; the 2008 Cluster Munitions Convention; and the 2013 Arms Trade Treaty.

The 1868 St. Petersburg Declaration[1]

Considering that the progress of civilization should have the effect of alleviating as much as possible the calamities of war.....The only legitimate object which States should endeavour to accomplish during war is to weaken the military forces of the enemy....This object would be exceeded by the employment of arms which uselessly aggravate the sufferings of disabled men, or render their death inevitable.

The 1907 Hague Convention (IV)
Respecting the Laws and Customs of War on Land

Until a more complete code of the laws of war has been issued, the high contracting Parties deem it expedient to declare that, in cases not included in the Regulations adopted by them, the inhabitants and the belligerents remain under the protection and the rule of the principles of the law of nations, as they result from the usages established among civilized peoples, from the laws of humanity, and the dictates of the public conscience.

---

1   The 1968 St. Petersburg Declaration has been regarded as the first major international agreement prohibiting the use of a particular weapon in warfare. The declaration renounced the use, in time of war, of explosive projectiles under 400 gm weight.

Regulations Respecting the Laws and Customs of War on Land

Article 1

The laws, rights, and duties of war apply not only to armies, but also to militia and volunteer corps, fulfilling the following conditions:

1. To be commanded by a person responsible for his subordinates;

2. To have a fixed distinctive emblem recognizable at a distance;

3. To carry arms openly; and

4. To conduct their operations in accordance with the laws and customs of war.

In countries where militia or volunteer corps constitute the army, or form part of it, they are included under the denomination "army".

Article 22

The right of belligerents to adopt means of injuring the enemy is not unlimited.

Article 23

Besides the prohibitions provided by special Conventions, it is especially prohibited:

(a) To employ poison or poisoned arms;

(b) To kill or wound treacherously individuals belonging to the hostile nation or army;

(c) To kill or wound an enemy who, having laid down arms, or having no longer means of defence, has surrendered at discretion;

(d) To declare that no quarter will be given;

(e) To employ arms, projectiles, or material of a nature to cause superfluous injury;

(f) To make improper use of a flag of truce, the national flag, or military ensigns and the enemy's uniform, as well as the distinctive badges of the Geneva Convention;

(g) To destroy or seize the enemy's property, unless such destruction or seizure be imperatively demanded by the necessities of war.

Article 25

The attack or bombardment of towns, villages, habitations or buildings which are not defended, is prohibited.

Article 26

The Commander of an attacking force, before commencing a bombardment, except in the case of an assault, should do all he can to warn the authorities.

Geneva Convention (I) for the Amelioration of the Condition of the Wounded and Sick in Armed Forces in the Field of August 12, 1949

Article 1

Respect for the Convention

The High Contracting Parties undertake to respect and to ensure respect for the present Convention in all circumstances.

Article 3

Conflicts not of an international character

In the case of armed conflict not of an international character occurring in the territory of one of the High Contracting Parties, each Party to the conflict shall be bound to apply, as a minimum, the following provisions:

Persons taking no active part in the hostilities, including members of armed forces who have laid down their arms and those placed hors de combat by sickness, wounds, detention, or any other cause, shall in all circumstances be treated humanely, without any adverse distinction founded on race, colour, religion or faith, sex, birth or wealth, or any other similar criteria.

Article 12

Protection and care

Members of the armed forces and other persons mentioned in the following Article, who are wounded or sick, shall be respected and protected in all circumstances.

They shall be treated humanely and cared for by the Party to the conflict in whose power they may be, without any adverse distinction founded on sex, race, nationality, religion, political opinions, or any other similar criteria. Any attempts upon their lives, or violence to their persons, shall be strictly prohibited; in particular, they shall not be murdered or exterminated, subjected to torture or to biological experiments; they shall not wilfully be left without medical assistance and care, nor shall conditions exposing them to contagion or infection be created.

Only urgent medical reasons will authorize priority in the order of treatment to be administered. Women shall be treated with all consideration due to their sex.

The Party to the conflict which is compelled to abandon wounded or sick to the enemy shall, as far as military considerations permit, leave with them a part of its medical personnel and material to assist in their care.

Article 13

Protected persons

The present Convention shall apply to the wounded and sick belonging to the following categories:

(1) Members of the armed forces of a Party to the conflict, as well as members of militias or volunteer corps forming part of such armed forces.

(2) Members of other militias and members of other volunteer corps, including those of organized resistance movements, belonging to a Party to the conflict and operating in or outside their own territory, even if this territory is occupied, provided that such militias or volunteer corps, including such organized resistance movements, fulfil the following conditions:

(a) that of being commanded by a person responsible for his subordinates;

(b) that of having a fixed distinctive sign recognizable at a distance;

(c) that of carrying arms openly;

(d) that of conducting their operations in accordance with the laws and customs of war.

(3) Members of regular armed forces who profess allegiance to a Government or an authority not recognized by the Detaining Power.

(4) Persons who accompany the armed forces without actually being members thereof, such as civil members of military aircraft crews, war correspondents, supply contractors, members of labour units or of services responsible for the welfare of the armed forces, provided that they have received authorization from the armed forces which they accompany.

(5) Members of crews, including masters, pilots and apprentices, of the merchant marine and the crews of civil aircraft of the Parties to the conflict, who do not benefit by more favourable treatment under any other provisions in international law.

(6) Inhabitants of a non-occupied territory, who on the approach of the enemy, spontaneously take up arms to resist the invading forces, without having had time to form themselves into regular armed units, provided they carry arms openly and respect the laws and customs of war.

Article 15

Search for casualties, Evacuation

At all times, and particularly after an engagement, Parties to the conflict shall, without delay, take all possible measures to search for and collect the wounded and sick, to protect them against pillage and ill-treatment, to ensure their adequate care, and to search for the dead and prevent their being despoiled.

Whenever circumstances permit, an armistice or a suspension of fire shall be arranged, or local arrangements made, to permit the removal, exchange and transport of the wounded left on the battlefield.

Geneva Convention (II) for the Amelioration of the Condition of Wounded, Sick and Shipwrecked Members of Armed Forces at Sea of August 12, 1949

Article 18

Search for casualties after an engagement

After each engagement, Parties to the conflict shall, without delay, take all possible measures to search for and collect the shipwrecked, wounded and sick, to protect them against pillage and ill-treatment, to ensure their adequate care, and to search for the dead and prevent their being despoiled.

Whenever circumstances permit, the Parties to the conflict shall conclude local arrangements for the removal of the wounded and sick by sea from a besieged or encircled area and for the passage of medical and religious personnel and equipment on their way to that area.

Protocol Additional to the Geneva Conventions of 12 August 1949, and relating to the Protection of Victims of International Armed Conflicts

(Protocol I), 1977[2]

Article 1

General principles and scope of application

1. The High Contracting Parties undertake to respect and to ensure respect for this Protocol in all circumstances.

2. In cases not covered by this Protocol or by other international agreements, civilians and combatants remain under the protection and authority of the

---

2   Adopted on 8 June 1977 by the Diplomatic Conference on the Reaffirmation and Development of International Humanitarian Law applicable in Armed Conflicts. Entry into force 7 December 1979, in accordance with Article 95.

principles of international law derived from established custom, from the principles of humanity and from the dictates of public conscience.

Article 10

Protection and care

1. All the wounded, sick and shipwrecked, to whichever Party they belong, shall be respected and protected.

2. In all circumstances they shall be treated humanely and shall receive, to the fullest extent practicable and with the least possible delay, the medical care and attention required by their condition. There shall be no distinction among them founded on any grounds other than medical ones.

Article 35

Basic rules

1. In any armed conflict, the right of the Parties to the conflict to choose methods or means of warfare is not unlimited.

2. It is prohibited to employ weapons, projectiles and material and methods of warfare of a nature to cause superfluous injury or unnecessary suffering.

3. It is prohibited to employ methods or means of warfare which are intended, or may be expected, to cause widespread, long-term and severe damage to the natural environment.

Article 36

New weapons

In the study, development, acquisition or adoption of a new weapon, means or method of warfare, a High Contracting Party is under an obligation to determine whether its employment would, in some or all circumstances, be prohibited by this Protocol or by any other rule of international law applicable to the High Contracting Party.

Article 40

Quarter

It is prohibited to order that there shall be no survivors, to threaten an adversary therewith or to conduct hostilities on this basis.

Article 41

Safeguard of an enemy hors de combat

1. A person who is recognized or who, in the circumstances, should be recognized to be hors de combat shall not be made the object of attack.

2. A person is hors de combat if:

   (a) He is in the power of an adverse Party;

   (b) He clearly expresses an intention to surrender; or

   (c) He has been rendered unconscious or is otherwise incapacitated by wounds or sickness, and therefore is incapable of defending himself; provided that in any of these cases he abstains from any hostile act and does not attempt to escape.

Article 42

Occupants of aircraft

1. No person parachuting from an aircraft in distress shall be made the object of attack during his descent.

2. Upon reaching the ground in territory controlled by an adverse Party, a person who has parachuted from an aircraft in distress shall be given an opportunity to surrender before being made the object of attack, unless it is apparent that he is engaging in a hostile act.

Article 48

Basic rule

In order to ensure respect for and protection of the civilian population and civilian objects, the Parties to the conflict shall at all times distinguish between the civilian population and combatants and between civilian objects and military objectives and accordingly shall direct their operations only against military objectives.

Article 51

Protection of the civilian population

1. The civilian population and individual civilians shall enjoy general protection against dangers arising from military operations. To give effect to this protection, the following rules, which are additional to other applicable rules of international law, shall be observed in circumstances.

2. The civilian population as such, as well as individual civilians, shall not be the object of attack. Acts or threats of violence the primary purpose of which is to spread terror among the civilian population are prohibited.

3. Civilians shall enjoy the protection afforded by this Section, unless and for such time as they take a direct part in hostilities.

4. Indiscriminate attacks are prohibited. Indiscriminate attacks are:

   (a) Those which are not directed at a specific military objective;

   (b) Those which employ a method or means of combat which cannot be directed at a specific military objective; or

   (c) Those which employ a method or means of combat the effects of which cannot be limited as required by this Protocol;

   and consequently, in each such case, are of a nature to strike military objectives and civilians or civilian objects without distinction.

5. Among others, the following types of attacks are to be considered as indiscriminate:

   (a) An attack by bombardment by any methods or means which treats as a single military objective a number of clearly separated and distinct military objectives located in a city, town, village or other area containing a similar concentration of civilians or civilian objects; and

   (b) An attack which may be expected to cause incidental loss of civilian life, injury to civilians, damage to civilian objects, or a combination thereof, which would be excessive in relation to the concrete and direct military advantage anticipated.

6. Attacks against the civilian population or civilians by way of reprisals are prohibited.

7. The presence or movements of the civilian population or individual civilians shall not be used to render certain points or areas immune from military operations, in particular in attempts to shield military objectives from attacks or to shield, favour or impede military operations. The Parties to the conflict shall not direct the movement of the civilian population or individual civilians in order to attempt to shield military objectives from attacks or to shield military operations.

8. Any violation of these prohibitions shall not release the Parties to the conflict from their legal obligations with respect to the civilian population and civilians, including the obligation to take the precautionary measures provided for in Article 57.

Article 52

General protection of civilian objects

1. Civilian objects shall not be the object of attack or of reprisals. Civilian objects are all objects which are not military objectives as defined in paragraph 2.

2. Attacks shall be limited strictly to military objectives. In so far as objects are concerned, military objectives are limited to those objects which by their nature, location, purpose or use make an effective contribution to military action and whose total or partial destruction, capture or neutralization, in the circumstances ruling at the time, offers a definite military of advantage.

3. In case of doubt whether an object which is normally dedicated to civilian purposes, such as a place of worship, a house or other dwelling or a school, is being used to make an effective contribution to military action, it shall be presumed not to be so used.

Article 57

Precautions in attack

1. In the conduct of military operations, constant care shall be taken to spare the civilian population, civilians and civilian objects.

2. With respect to attacks, the following precautions shall be taken:

(a) Those who plan or decide upon an attack shall:

(i) Do everything feasible to verify that the objectives to be attacked are neither civilians nor civilian objects and are not subject to special protection but are military objectives within the meaning of paragraph 2 of Article 52 and that it is not prohibited by the provisions of this Protocol to attack them;

(ii) Take all feasible precautions in the choice of means and methods of attack with a view to avoiding, and in any event to minimizing, incidental loss of civilian life, injury to civilians and damage to civilian objects;

(iii) Refrain from deciding to launch any attack which may be expected to cause incidental loss of civilian life, injury to civilians, damage to civilian objects, or a combination thereof, which would be excessive in relation to the concrete and direct military advantage anticipated;

(b) An attack shall be cancelled or suspended if it becomes apparent that the objective is not a military one or is subject to special protection or that the attack may be expected to cause incidental loss of civilian life, injury to civilians, damage to civilian objects, or a combination thereof, which would be excessive in relation to the concrete and direct military advantage anticipated;

(c) Effective advance warning shall be given of attacks which may affect the civilian population, unless circumstances do not permit.

3. When a choice is possible between several military objectives for obtaining a similar military advantage, the objective to be selected shall be that the attack on which may be expected to cause the least danger to civilian lives and to civilian objects.

4. In the conduct of military operations at sea or in the air, each Party to the conflict shall, in conformity with its rights and duties under the rules of international law applicable in armed conflict, take all reasonable precautions to avoid losses of civilian lives and damage to civilian objects.

5. No provision of this Article may be construed as authorizing any attacks against the civilian population, civilians or civilian objects.

Article 87

Duty of commanders

1. The High Contracting Parties and the Parties to the conflict shall require military commanders, with respect to members of the armed forces under their command and other persons under their control, to prevent and, where necessary, to suppress and to report to competent authorities breaches of the Conventions and of this Protocol.

2. In order to prevent and suppress breaches, High Contracting Parties and Parties to the conflict shall require that, commensurate with their level of responsibility, commanders ensure that members of the armed forces under their command are aware of their obligations under the Conventions and this Protocol.

3. The High Contracting Parties and Parties to the conflict shall require any commander who is aware that subordinates or other persons under his control are going to commit or have committed a breach of the Conventions or of this Protocol, to initiate such steps as are necessary to prevent such violations of the Conventions or this Protocol, and, where appropriate, to initiate disciplinary or penal action against violators thereof.

Article 91

Responsibility

A Party to the conflict which violates the provisions of the Geneva Conventions or of this Protocol shall, if the case demands, be liable to pay compensation. It shall be responsible for all acts committed by persons forming part of its armed forces.

The 1998 Rome Statute of the International Criminal Court [3]

Article 28

Responsibility of commanders and other superiors

In addition to other grounds of criminal responsibility under this Statute for crimes within the jurisdiction of the Court:

(a) A military commander or person effectively acting as a military commander shall be criminally responsible for crimes within the jurisdiction of the Court committed by forces under his or her effective command and control, or effective authority and control as the case may be, as a result of his or her failure to exercise control properly over such forces, where:

    (i) That military commander or person either knew or, owing to the circumstances at the time, should have known that the forces were committing or about to commit such crimes; and

    (ii) That military commander or person failed to take all necessary and reasonable measures within his or her power to prevent or repress their commission or to submit the matter to the competent authorities for investigation and prosecution.

(b) With respect to superior and subordinate relationships not described in paragraph (a), a superior shall be criminally responsible for crimes within the jurisdiction of the Court committed by subordinates under his or her effective authority and control, as a result of his or her failure to exercise control properly over such subordinates, where:

    (i) The superior either knew, or consciously disregarded information which clearly indicated, that the subordinates were committing or about to commit such crimes;

    (ii) The crimes concerned activities that were within the effective responsibility and control of the superior; and

    (iii) The superior failed to take all necessary and reasonable

---

3   The Statute entered into force on 1 July 2002.

measures within his or her power to prevent or repress their commission or to submit the matter to the competent authorities for investigation and prosecution.

The 2014 Arms Trade Treaty

Principles

- Respecting and ensuring respect for IHL in accordance with, inter alia, the Geneva Conventions of 1949, and respecting and ensuring respect for human rights in accordance with, inter alia, the Charter of the United Nations and the Universal Declaration of Human Rights.

- The responsibility of all States, in accordance with their respective international obligations, to effectively regulate the international trade in conventional arms, and to prevent their diversion, as well as the primary responsibility of all States in establishing and implementing their respective national control systems.

Article 2

Scope

1. This Treaty shall apply to all conventional arms within the following categories: (a) Battle tanks; (b) Armoured combat vehicles; (c) Large-calibre artillery systems; (d) Combat aircraft; (e) Attack helicopters; (f) Warships; (g) Missiles and missile launchers; and (h) Small arms and light weapons.

2. For the purposes of this Treaty, the activities of the international trade comprise export, import, transit, trans-shipment and brokering, hereafter referred to as "transfer".

3. This Treaty shall not apply to the international movement of conventional arms by, or on behalf of, a State Party for its use provided that the conventional arms remain under that State Party's ownership.

Article 3

Ammunition/Munitions

Each State Party shall establish and maintain a national control system to regulate the export of ammunition/munitions fired, launched or delivered by the conventional arms covered under Article 2 (1), and shall apply the provisions of Article 6 and Article 7 prior to authorizing the export of such ammunition/munitions.

Article 4

Parts and Components

Each State Party shall establish and maintain a national control system to regulate the export of parts and components where the export is in a form that provides the capability to assemble the conventional arms covered under Article 2 (1) and shall apply the provisions of Article 6 and Article 7 prior to authorizing the export of such parts and components.

Article 6

Prohibitions

1. A State Party shall not authorize any transfer of conventional arms covered under Article 2 (1) or of items covered under Article 3 or Article 4, if the transfer would violate its obligations under measures adopted by the UN Security Council acting under Chapter VII of the Charter of the United Nations, in particular arms embargoes.

2. A State Party shall not authorize any transfer of conventional arms covered under Article 2 (1) or of items covered under Article 3 or Article 4, if the transfer would violate its relevant international obligations under international agreements to which it is a Party, in particular those relating to the transfer of, or illicit trafficking in, conventional arms.

3. A State Party shall not authorize any transfer of conventional arms covered under Article 2 (1) or of items covered under Article 3 or Article 4, if it has knowledge at the time of authorization that the arms or items would be used in the commission of genocide, crimes against humanity, grave breaches of the Geneva Conventions of 1949, attacks directed against civilian objects or civilians protected as such, or other war crimes as defined by international agreements to which it is a Party.

Article 7

Export and Export Assessment

1. If the export is not prohibited under Article 6, each exporting State Party, prior to authorization of the export of conventional arms covered under Article 2 (1) or of items covered under Article 3 or Article 4, under its jurisdiction and pursuant to its national control system, shall, in an objective and non-discriminatory manner, taking into account relevant factors, including information provided by the importing State in accordance with Article 8 (1), assess the potential that the conventional arms or items:

    (a) would contribute to or undermine peace and security;

    (b) could be used to: (i) commit or facilitate a serious violation of IHL; (ii) commit or facilitate a serious violation of international human rights law; (iii) commit or facilitate an act constituting an offence under international conventions or protocols relating to terrorism to which the exporting State is a Party; or (iv) commit or facilitate an act constituting an offence under international conventions or protocols relating to transnational organized crime to which the exporting State is a Party.

2. The exporting State Party shall also consider whether there are measures that could be undertaken to mitigate risks identified in paragraph 1 (a) or (b), such as confidence-building measures or jointly developed and agreed programmes by the exporting and importing States.

3. If, after conducting this assessment and considering available mitigating measures, the exporting State Party determines that there is an overriding risk of any of the negative consequences in paragraph 1, the exporting State Party shall not authorize the export.

4. The exporting State Party, in making this assessment, shall take into account the risk of the conventional arms covered under Article 2 (1) or of the items covered under Article 3 or Article 4 being used to commit or facilitate serious acts of gender-based violence or serious acts of violence against women and children.

5. Each exporting State Party shall take measures to ensure that all

authorizations for the export of conventional arms covered under Article 2 (1) or of items covered under Article 3 or Article 4 are detailed and issued prior to the export.

6. Each exporting State Party shall make available appropriate information about the authorization in question, upon request, to the importing State Party and to the transit or trans-shipment States Parties, subject to its national laws, practices or policies.

7. If, after an authorization has been granted, an exporting State Party becomes aware of new relevant information, it is encouraged to reassess the authorization after consultations, if appropriate, with the importing State.

Appendix B

Draft

Protocol on Prohibitions on the Development, Transfer and Use of Lethal Autonomous Weapon Systems

(Protocol VI)

Article 1

Material scope of application

This protocol relates to the development and use on land, air, space or sea of the lethal autonomous weapon systems (LAWS) and other devices defined herein, but does not apply to the use of autonomous weapon systems wherein the decision to use the lethal force is taken by a human operator of such systems.

Article 2

Definitions

For the purpose of this protocol:

1. 'Lethal autonomous weapon systems' means weapon systems which select and engage targets without a human operator; where lethal force is directed at human beings.

2. 'Autonomous' means that the technological capability of the weapon systems is such that they can function without any intervention by a human operator.

3. 'Lethal' means capable of causing death or serious injuries.

4. 'Human operator' means a responsible person in the military chain of command who is authorized to use autonomous weapon systems in an armed conflict.

5. 'Weapon systems' means the actual weapons, as well as the storage, transportation and delivery mechanisms of such systems.

Article 3

Prohibition on the development and use of lethal autonomous weapon systems

Developing, acquiring, transferring and trading in a lethal autonomous weapon system, or deploying such weapon systems in an armed conflict is prohibited.

Article 4

Deployment and the use of autonomous weapon systems

In the deployment and use of autonomous weapon systems, the High Contracting Parties shall take all feasible measures to ensure that the decision to use lethal force during the employment of such weapon systems is taken by a human operator.

Bibliography

Abbott Kenneth W and Duncan Snidal, Hard and Soft Law in International Governance, *International Organization*, Vol. 54, No. 3, Summer 2000, p. 421–456.

Advancing the Debate on Killer Robots: 12 Key Arguments for a Preemptive Ban on Fully Autonomous Weapons, Human Rights Watch Report, May 2014.

Ajda Hosseini Ghasemi, Semi-Autonomous Weapon Systems in International Humanitarian Law, Unpublished Thesis, Lund University, 2014.

A Guide to the Legal Review of New Weapons, Means and Methods of Warfare: Measures to Implement Article 36 of Additional Protocol I of 1977, *International Review of the Red Cross*, Vol. 88, No. 864, December 2006, p. 931-956.

A Roadmap for US Robotics: From Internet to Robotics, 2013, available at: https://robotics-vo.us/sites/default/files/2013%20Robotics%20 Roadmap-rs.pdf.

Akgijl Aziz, Artificial Intelligence: Military Applications, Turkish Military Academy.

Allenby Braden R., Are new technologies undermining the laws of war? *Bulletin of the Atomic Scientists*, 2014, Vol. 70 (1), 201, p. 21–31.

Alston Philip, Lethal Robotic Technologies: The Implications for Human Rights and International Humanitarian Law, *Journal of Law, Information and Science*, Vol. 21 (2), 2012.

Altmann Jurgen, Peter Asaro, Noel Sharkey and Robert Sparrow, Armed

Military Robots (Editorial), *Ethics Inf Technol*, Vol. 15, 2013, p. 73-76.

Altmann Jurgen, 'Preventive Arms Control for Uninhabited Military Vehicles', in Capurro R. and Nagenborg M. (eds.). 2009. *Ethics and Robotics*, AKA Verlag Heidelberg, p. 69-82.

Altmann Jurgen, Arms control for armed uninhabited vehicles: an ethical issue, *Ethics Inf Technol*, Vol. 15, (2013), p. 137–152.

Andersson Cecelia, Killer Robots: Autonomous Weapons and Their Compliance with IHL, Unpublished Thesis, Faculty of Law, Lund University, 2014.

Anderson Kenneth, Comparing the Strategic and Legal Features of Cyberwar, Drone Warfare, and Autonomous Weapon Systems, 27 February 2015, Policy Comment, Hoover Institution, Stanford University.

Anderson, Kenneth, Daniel Reisner and Matthew Waxman, Adapting the Law of Armed Conflict to Autonomous Weapon Systems, *International Law Studies*, Vol. 90, 2014, p. 386-411.

Anderson Kenneth and Matthew Waxman, Law and Ethics for Autonomous Weapon Systems: Why a Ban Won't Work and How the Laws of War Can, American University Washington College of Law, Research Paper No. 2013-11, Columbia Public Law Research Paper, p. 1-33.

Anderson Kenneth and Matthew Waxman, Law and Ethics for Robot Soldiers, *Policy Review*, December 2012/January 2013, p. 35-49.

Anderson Michael and Anderson Susan Leigh, Machine Ethics: Creating an Ethical Intelligent Agent, *AI Magazine*, Vol. 28, No. 4 (2007), p. 15-26.

Anderson, Susan Leigh, Asimov's "three laws of robotics" and machine metaethics, *AI & Society*, Vol. 22(4), 2008, p. 477–493.

Anthony Ian and Chris Holland, The Governance of Autonomous Weapons, *SIPRI Yearbook 2014*, p. 423-413.

Arkin, Ronald C., *Ethical Robots in Warfare*, Georgia Institute of Technology, Mobile Robot Lab, College of Computing, Atlanta, p. 1-5.

Arkin Ronald C. 2009, *Governing Lethal Behaviour in Autonomous Robots*, USA: Chapman & Hall/CRC.

Arkin, Ronald C., The Case for Ethical Autonomy in Unmanned Systems, *Journal of Military Ethics*, Vol. 9, No. 4, 2010, p. 332-341.

Arkin Ronald, Lethal Autonomous Systems and the Plight of the Non-combatant, *AISB Quarterly*, No. 137, July 2013, p. 1-8.

Arkin, Ronald. 2009. *Governing Lethal Behavior in Autonomous Robots*, Boca Raton: Travis and Francis Group.

Arkin, Ronald C. and Patrick Ulam, Overriding Ethical Constraints in Lethal Autonomous Systems, Technical Report GIT-MRL-12-01, Georgia Institute of Technology, Atlanta, USA.

Article 36, Killer Robots: UK Government Policy on Fully Autonomous Weapons, April 2013.

Asaro Peter, On banning autonomous weapon systems: human rights, automation, and the dehumanization of lethal decision-making, *International Review of the Red Cross*, Vol. 94, No. 886, Summer 2012, p. 687-709.

Autonomous Weapon Systems: Five Key Human Rights Issues For Consideration, Amnesty International Publications, Index: ACT 30/1401/2015.

Autonomous weapon systems: Technical, military, legal and humanitarian aspects, Proceedings of an Expert Meeting, ICRC, Geneva, Switzerland, 26-28 March 2014.

Autonomy in Weapon Systems, The US Department of Defence, Directive No. 3000.09, 21 November 2012.

Backstrom Alan and Henderson Ian, New Capabilities in Warfare: An Overview of contemporary technological developments and associated legal and engineering issues in Article 36 weapons review, *International Review of the Red Cross*, Vol. 94, No. 886, Summer 2012, p. 483-514.

Ballesteros Stefano, Redefining War: Implementation of Autonomous Weapons Systems, available at: http://fanobelmont.com/assets/

redefining-war.pdf.

Beard Jack M., Autonomous Weapons and Human Responsibilities, *Georgetown Journal of International Law*, Vol. 45, 2014, p. 617-681.

Beardsley Steven, NATO researchers test underwater drones in Norway, 6 May 2015, available at: http://www.stripes.com/nato-researchers-test-underwater-drones-in-norway-1.344499.

Berkowitz Bruce, Sea Power in the Robotic Age, ISSUES in Science and Technology, 2015, available at; http://issues.org/30-2/bruce-2/.

Bertolini Andrea and Palmerini Erica, Regulating Robotics: A Challenge for Europe, Workshop on Upcoming Issues of EU Law, September 2014, p. 169-200.

Bieri Matthias and Marcel Dickow, Lethal Autonomous Weapons Systems: Future Challenges, Centre for Security Studies Analyses, No. 164, November 2014, p. 1-4.

Bills Gwendelynn, Laws unto Themselves: Controlling the Development and Use of Lethal Autonomous Weapons Systems, *The George Washington Law Review,* December 2014, Vol. 83 No. 1, p. 176-208.

Blank Laurie R., Extending Positive Identification From Person to Places: Terrorism, Armed Conflict and the Identification of Military Objectives, *Utah Law Review*, 2013, No. 5, p. 1227-1261.

Blidberg D. Richard, The Development of Autonomous Underwater Vehicles (AUV); A Brief Summary, available at: http://www.ausi.org/publications/ICRA_01paper.pdf.

Bolton Matthew and Wim Zwijnenburg, Future-proofing is Never Complete: Ensuring the Arms Trade Treaty Keeps Pace with New Weapons Technology, ICRAC Working Paper No. 1, October 2013.

Boothby Bill, Autonomous attack—Opportunity or Spectre, *Yearbook of International Humanitarian Law*, Vol. 16, 2013, p. 71-87.

Boothby William H., Autonomous systems: Precautions in attacks, *International Humanitarian Law and New Weapon Technologies*, Proceedings of 34th Round Table on Current Issues of International

Humanitarian Law, Sanremo, Italy, 8-10 September 2011, p. 119-124.

Boothby William, Some legal challenges posed by remote attack, *International Review of the Red Cross*, Vol. 94, N0. 886, Summer 2012, p. 579-596.

Boothby William H. 2014. 'The Legal Challenges of New Technologies: An Overview', in Nasu H. and R. McLaughlin (eds.), *New Technologies and the Law of Armed Conflict*, TMC Asser Press, p. 21-28.

Boothby William H. 2014. Conflict Law-The Influence of New Weapon Technology, Human Rights and Emerging Actors, TMC Asser Press.

Boothby William H., 'The Legal Challenges of New Technologies: An Overview', in Nasu Hitoshi and Robert McLaughlin (ed.). 2014. *New Technologies and the Law of Armed Conflict*, The Hague: TMC Asser Press, p. 21-29.

Boulanin Vincent, Implementing Article 36 Weapon Reviews in the Light of Increasing Autonomy in Weapon Systems, SIPRI Insight on Peace and Security, No. 2015/1, November 2015.

Boyhan David M., Autonomous Weapons Proliferation, available at: https://itp.nyu.edu/classes/foti-fall2011/files/2011/11/autonomous-weapons-dboyhan.pdf.

Bradshaw Jeffrey M., Robert R. Hoffman, Matthew Johnson, and David D. Woods, The Seven Deadly Myths of "Autonomous Weapons", May/June 2013, *Human - Centred Computing*, May/June 2013, p. 57.

Brannen Samuel, *Sustaining the US lead in Unmanned Systems: Military and Homeland Considerations through 2025*, February 2014, Centre for Strategic and International Studies, Washington.

Bringsjord Selmer and Joshua Taylor, 'The Divine-Command Approach to Robot Ethics', in Lin Patrick, Keith Abney, and Bekey George A, (ed). 2012. *Robotic Ethics: The Ethical and Social Implications of Robotics*, Cambridge: MIT Press, p. 85-108.

Brooks Rosa, In Defence of Killer Robots, *Foreign Policy*, 18 May 2015.

Brown Gary D. and Andrew O. Metcalf, Easier Said Than Done: Legal

Reviews of Cyber Weapons, Journal of National Security Law and Policy, Vol. 7, 2014, p. 115-138.

Byrnes, Capt Michael W., Nightfall: Machine Autonomy in Air-to-Air Combat, *Air & Space Power Journal*, May-June 2014, p. 48-75.

Calo Ryan, Robotics and the New Cyber Law, available at: http://robots. law.miami.edu/2014/wp-content/uploads/2013/06/Calo-Robotics-and-the-New-Cyberlaw.pdf.

Campaign to Stop Killer Robots, Report on Activities, Geneva, May 2014.

Carafano James Jay, Autonomous Military Technology: Opportunities and Challenges for Policy and Law, Backgrounder No 2932, 6 August 2014.

Carpenter Charli, Beware the Killer Robots, Inside the debate over Autonomous Weapons, *Foreign Affairs*, 3 July 2013.

Casey-Maslen Stuart, Pandora Box? Drone strikes under *jus ad bellum, jus in bello* and international human rights law, *International Review of the Red Cross*, Vol. 94, N0. 886, Summer 2012, p. 597-626.

Cassese Antonio, The Martens Clause: Half a Loaf or simply Pie in the Sky? *European Journal of International Law*, Vol. 11, No. 1, 2000, p. 187-216.

Chayes Antonia, Rethinking Warfare: The Ambiguity of Cyber Attacks, *Harvard National Security Journal*, Vol. 6, 2015, p. 474-519.

Coeckelbergh Mark, Can we trust robots? *Ethics Inf Technol*, Vol. 14, 2012, p. 53–60.

Coeckelbergh Mark, War from a Distance: The Ethics of Killer Robots, 14 June 2014, available at: http://www.e-ir.info/2014/06/16/war-from-a-distance-the-ethics-of-killer-robots/.

Coleman Stephen, Possible Ethical Problems with Military Use of Non-Lethal Weapons, *Case W. Res. J. Int'l L.*, Vol. 47 (1), 2015, p. 185-199.

Coleman Stephen, 'Ethical Challenges of New Military Technologies', in Nasu Hitoshi and Robert McLaughlin (ed.). 2014. *New Technologies and the Law of Armed Conflict*, The Hague: TMC Asser Press, p. 29-42.

Copeland Damian P., 'Legal Review of New Technology Weapon', in Nasu

Hitoshi and Robert McLaughlin (ed.). 2014. *New Technologies and the Law of Armed Conflict*, The Hague: TMC Asser Press, p. 43-55.

Corn Geoffrey S., Autonomous Weapon Systems: Managing the Inevitability of "Taking the Man out of the Loop", available at: http://ssrn.com/abstract=2450640.

Corn Geoffrey S., Mixing Apples with Hand Grenades: The Logical limit of Applying Human Rights Norms to Armed Conflict, available at: http://ssrn.com/abstract=1511954.

Country Statements on Killer Robots, Compilation by the Campaign to Stop Killer Robots, March 2014.

Cowan Thomas H., A Theoretical, Legal and Ethical Impact of Robots on Warfare, Student Project Report, 2007, The US Army War College.

Crootof Rebecca, The Killer Robots are Here: Legal and Policy Implications, *Cardozo Law Review*, Vol. 36, 2015, p. 1837-1915.

Crootof Rebecca, The varied Law of Autonomous Weapon Systems, available at: http://ssrn.com/abstract=2569322.

Dan Saxon (ed). 2013. *International Humanitarian Law and Changing Technology of War*, Martin Nijhoff.

Darren M. Stewart, New Technology and the Law of Armed Conflict, *International Law Studies*, Vol. 87, 2011, p. 271-298.

Deborah G. Johnson and Noorman, Merel E., Responsibility Practices in Robotic Warfare, *Military Review*, May-June 2014, p. 12-22.

Deputy Samuel N., Counterinsurgency and Robots: Will the means undermine the end, Research paper submitted to the Naval War College, Newport, report dated 04 May 2009.

Doare Ronan, Jean-Paul Hanon and Gerard de Boisboissel (ed). 2014. *Robots on the Battlefield: Contemporary Perspectives and Implications for the Future*, Combat Studies Institute Press, US Army Combined Arms Centre, Kansas.

Dill Janina, The 21st-Century Belligerent's Trilemma, *The European Journal of International Law*, Vol. 26, No. 1, p. 83-108.

Docherty Bonnie, The Human Rights Implications of 'Killer Robots', *Jurist*, 10 June 2014.

Doermann Knut, Obligations of International Humanitarian Law, *Military and Strategic Affairs*, Vol. 4, No. 2, September 2012, p. 11-23.

Droege Cordula, Get off my cloud: Cyber warfare, IHL, and the protection of civilians, *International Review of the Red Cross*, Vol. 94, No. 886, Summer 2012, p. 533-578.

Ekelhof Merel and Miriam Struyk, Deadly Decisions: Eight Objections to Killer Robots, PAX, The Netherlands, 2014.

Espada Cesareo Gutierrez and Maria Jose Cervell Hortal, Autonomous weapons systems, drones and international law, Revista del Instituto Espanol de Estudios Estrategicos, No. 2 of 2013, p. 1-30.

Evans Tyler D., At War with the Robots: Autonomous Weapons Systems and the Martens Clause, *Hofstra Law Review*, Vol. 41, 2013, p. 697-733.

Fielding Marcus, Robotics in Future Land Warfare, *Australian Army Journal*, Vol. III, No. 2, p. 99-108, Winter 2006.

Foster Melisa and Haden-Pawlowsky Virgil, Regulating Robocop: The need for International Governance Innovation in Drone Development and AWS Development and Use, CIGI Junior Fellows, Policy Brief, No. 18, October 2014.

Foy, James, Autonomous Weapons Systems: Taking the Human out of International Humanitarian Law, *Dalhousie Journal of Legal Studies*, Vol. 23, 2014, p. 47-70.

Framing Discussions on the Weaponization of Increasingly Autonomous Technologies, The United Nations Institute for Disarmament Research (UNIDIR), 2014, p. 14.

Francois Camille, Robots, War, and Society: How robotics will alter the modern battlefield, and American culture, *Defence Dossier*, February 2015, Issue 13, p. 11-15.

Frowe Helen. 2011. *The Ethics of War and Peace*, London: Routledge.

Fry James, D., The XM25 Individual Semi-automatic Airburst Weapon

System and International Law, *UNSW Law Journal*, Vol. 36, No. 2, p. 682-710.

Galliott Jai. 2015. Military Robots: Mapping the Moral Landscape, UK: Ashgate Publishing Ltd.

Garcia Denise, Humanitarian Security Regimes, *International Affairs, Vol.* 91, No. 1, 2015, p. 55–75.

Garcia Denise, The case against killer robots: why the United States should ban them, *Foreign Affairs*, May 2014. Available at: https://www. foreignaffairs.com/articles/united-states/2014-05-10/case-against-killer-robots.

Gary E. Marchant, et. al., International Governance of Automated Military Robots, *The Columbia Science and Technology Law Review*, Vol. XII, 2011, p. 272-315.

Geiss Robin, The International-Law Dimension of Autonomous Weapons Systems, October 2015, International Policy Analysis, Germany.

Geser Hans, Modest Prospective for Military Robots in Today's Asymmetric Wars, 2011, available at: http://www.geser.net/internat/t_hgeser8.pdf.

Gogarty Brendan and Meredith Hagger, The Laws of Man over Vehicles Unmanned: The Legal Response to Robotic Revolution on Sea, Land and Air, *Journal of Law, Information and Science*, Vol. 19, 2008, p. 73-145.

Goldsmith D.A., Robots in the Battlespace: Moral and Ethical Considerations in the Use of Autonomous Mechanical Combatants, Canadian Forces College, JCSP 34.

Goose Steve, The Need for a Preemptive Prohibition on Fully Autonomous Weapons, *Ethics and Armed Forces*, Issue 2014/1, p. 11-18.

Graff Bradley, Branch Jason, Gianetti Michael and Alotaibi Fawaz, Weapon Evolution: The Birth of Lethal Autonomous Robotics Systems, *Campaigning: The Journal of the Joint Forces Staff College*, Spring 2014, p. 94-106.

Grau Christopher, There is no 'I' in 'Robot': Robotic Utilitarians and Utilitarian Robots, 2005, American Association for Artificial Intelligence.

Gray Jeff, If a Robot Kills Someone, who is to blame? Available at: http://www.theglobeandmail.com/news/world/if-a-robot-kills-someone-who-is-to-blame/article23996250/.

Gros Florian, Tessier C. and Pichevin T, Ethics and Authority Sharing for Autonomous Armed Robots, available at: http://ceur-ws.org/Vol-885/paper1.pdf.

Groves Steven, The US Should Oppose the UN's Attempt to Ban Autonomous Weapons, The Heritage Foundation, 5 March 2015.

Grut, Chantal, The Challenge of Autonomous Lethal Robotics to International Humanitarian Law, *Journal of Conflict & Security Law*, Vol. 18, No. 1, 2013, p. 5–23.

Gubrud Mark, Stopping Killer Robots, *Bulletin of the Atomic Scientists*, Vol. 70, No. 1, 2014, p. 32-42.

Gubrud Mark and Altmann Jurgen, Compliance Measures for an Autonomous Weapons Convention, International Committee for Robots Arm Control (ICRAC) Working Paper No.2. p. 8.

Guetlein Maj Mike, Lethal Autonomous Weapons: Ethical and Doctrinal Implications, Paper submitted to the Naval War College in partial satisfaction of the requirements of the Department of Joint Military Operations, 2005.

Gulpers Liesbeth, Encouraging Moral Military Behaviour Through the Infrastructural Design of the Military Organization, Proceedings of the 16th Annual Working Conference, 2010.

Haecker Ryan, Chivalry in the Age of Autonomous Weapons, Part I and II, available at: http://transhumantraditionalism.blogspot.com/2008/03/chivalry-in-age-of-autonomousweapons.html.

Haas Michael Carl, Autonomous Weapons Systems: The Military's Smartest Toys? 20 November 2014, available at:http://nationalinterest.org/feature/autonomous-weapon-systems-the-militarys-smartest-toys-11708.

Hammond Daniel N., Autonomous Weapons and the Problem of State Accountability, *Chicago Journal of International Law*, Vol. 15, No. 2, 2015, p. 552-687.

Haruna A Lawan, Laminu Bukar and Babagana Karumi, Principle of Distinction in Armed Conflict: An Analysis of the Legitimacy of 'Combatants and Military Objectives' As a Military Target, *International Journal of Humanities and Social Science Invention*, Vol. 3, Issue 3, March 2014, p. 15-24.

Hattan Titus, Lethal Autonomous Robots: Are They Legal under International Human Rights and Humanitarian Law? *Nebraska Law Review*, Vol. 93, No.4, 2015, p. 1035-1057.

Hauptman Allyson, Autonomous Weapons and the Law of Armed Conflict, *Military Law Review*, Vol. 218, Winter 2013, p. 170-197.

Henderson Ian, Jordan den Dulk and Angeline Lewis, Emerging Technology and Perfidy in Armed Conflict, *International Law Studies*, Vol. 91, 2015, p. 468-485.

Henderson Ian. 2010. *The Contemporary Law of Targeting*, International Humanitarian Law Series, Vol. 25, Leiden: Martinus Nijhoff Publishers.

Herbach, Jonathan David, Into the Caves of Steel: Precaution, Cognition and Robotic Weapon Systems Under the International Law of Armed Conflict, *Amsterdam Law Forum*, Vol. 4, 2012, p. 3-20.

Hew Patrick, The Blind Spot in Robot-Enabled Warfare, *Australian Army Journal*, Vol. VII, No. 2, p. 45-56.

Heyns Christof, Report of the Special Rapporteur on extrajudicial, summary or arbitrary executions, UNGA Doc A/HRC/23/47 dated 9 April 2013.

Heyns Christof, Autonomous weapons systems and human rights law, Presentation made at the informal expert meeting organized by the state parties to the Convention on Certain Conventional Weapons 13 – 16 May 2014, Geneva, Switzerland.

Hoffman Robert R., Hawley John K. and Jeffrey M. Bradshaw, Myths of Automation Part 2, Some Very Human Consequences, *IEEE Computer*

Society, 2014, p. 82-85.

Horowitz Michael C. and Paul Scharre, An Introduction to Autonomy in Weapon Systems, Working Paper, The Centre for a New American Security, February 2015.

Horowitz Michael C. and Paul Scharre, Meaningful Human Control in Weapon Systems: A Primer, Working Paper, The Centre for a New American Security, March 2015, p. 16.

Horowitz Michael C. and Paul Scharre, The Morality of Robotic War, *The New York Times*, 26 May 2015.

Howlader Daniel and James Giordano, Advanced Robotics: Changing the Nature of War and Thresholds and Tolerance for Conflict - Implications for Research and Policy, *The Journal of Philosophy, Science & Law*, Vol. 13 (2013), p. 1-19.

Ian Henderson, 'Autonomous Weapons Debate: Three Issues for Consideration', *Just Security*, 13 June 2014.

Jacobsson Marie, Modern Weaponry and Warfare: The Application of Article 36 of Additional Protocol I by Governments, *International Law Studies*, Vol. 82, 2006, p. 183-191.

Jeferies Adrianne, Should a Robot Decide When to Kill? The ethics of war machines, 28 January 2014, available at: http://www.theverge. com/2014/1/28/5339246/war-machines-ethics-of-robots-on-the-battlefield.

Jensen Eric Talbot, The Future of the Law of Armed Conflict: Ostriches, Butterflies, and Nanobots, *Michigan Journal of International Law*, Vol. 35, No. 2, 2014, p. 253-317.

Jensen Eric Talbot, Future War, Future Law, *Minn J International Law*, Vol. 22, Summer 2013, p. 282-322.

Johnson Aaron and Axinn Sidney, The Morality of Autonomous Robots, *Journal of Military Ethics*, Vol. 12, Issue 2, August 2013.

Johnson Deborah G. and Merel E. Noorman, Responsibility Practices in Robotic Warfare, *Military Review*, May-June 2014, p. 12-21.

Kadu R.A., et. al., Wireless Control and Monitoring of Robotic Arm (SWORDS), *International Journal of Computer Technology and Electronics Engineering (IJCTEE)*, Vol. 2, Issue 1, p. 28-38.

Kastan Benjamin, Autonomous Weapons Systems: A Coming Legal "Singularity"? *Journal of Law, Technology and Policy*, Vol. 2013, No. 1, p. 45-82.

Kellenberger, Jakob, keynote address, International Humanitarian Law and New Weapon Technologies 34th round Table on current issues of IHL, San Remo, 8-10 September 2011, *International Review of the Red Cross*, Vol. 94, No. 886, Summer 2012, p. 809-822.

Kernaghan K., The Rights and Wrongs of Robotics: Ethics and Robotics in Public Organizations, *Canadian Public Administration*, Vol. 57, No. 4, December 2014, p. 485-506.

Killer Robots: UK Government Policy on Fully Autonomous Weapons, April 2013.

Koch Bernhard, Of Men and Machine. What Does the Robotization of the Military means from an Ethical Perspective? *Ethics and the Armed Forces*, Issue 2014/1, p. 19-22.

Kodar Erki, Applying the Law of Armed Conflict from the Martens Clause to Additional Protocol I, ENDC Proceedings, Vol. 15, 2012, p. 107-132.

Korybko Andrew, How the Pentagon Plans to Defeat Eurasia and Roll out "Robotic Warfare, 17 June 2015. Available at: http://www.globalresearch.ca/the-pentagons-robotic-warfare-arsenal/5456848.

Kovach Christopher M., Beyond *Skynet*: Reconciling Increased Autonomy in Computer-Based Weapon Systems with the Laws of War, *Air Force Law Review*, Vol. 71, 2014, p. 231-277.

Krause Keith, 'Leashing the dogs of war: arms control from sovereignty to governmentality', *Contemporary Security Policy*, Vol. 32, No. 1, April 2011, p. 20–39.

Kreit Alex, Vicarious Criminal Liability and the Constitutional Dimensions of Pinkerton, *American University Law Review*, Vol. 57, No. 3, 2008, p. 585-639.

Krishnan Armin. 2009. *Killer Robots: Legality and Ethicality of Autonomous Weapons*, USA: Ashgate.

Krishnan Armin, Robots, Soldiers, & Cyborgs: The Future of Warfare, 24 October 2013, available at: http://footnote1.com/robots-soldiers-cyborgs-the-future-of-warfare/.

Krupiy Tetyana, Of Souls, Spirit and Ghosts: Transposing the Application of the Rules of Targeting to Lethal Autonomous Robots, *Melbourne Journal of International Law*, Vol. 16, 2015, p. 1-58.

Kukita Minao, The difference between artificial intelligence and artificial morality, ICAE 2014, Hokkaido University, Sapporo, 31 October 2014, p. 7.

Kyyronen Vesa, Machines Making Decisions: The Applicability of State Responsibility Doctrine in the Case of Autonomous Systems, Unpublished thesis, University of Helsinki.

Latiff Robert and Howard Don, Ethical, Legal, and Societal Implications of New Weapons Technologies, available at: http://reilly.nd.edu/assets/169560/elsi_in_weapons_r_d_briefingbook_final_29april.pdf.

Lawand Kathleen, Fully autonomous weapon systems, available at: https://www.icrc.org/eng/resources/documents/statement/2013/09-03-autonomous-weapons.htm.

Legal Support to Military Operations, US Department of Defence, Joint Publication 1-04, 17 August 2011.

Leveringhaus Alex and Gilles Giacca, *Robo-Wars: The Regulation of Robotic Weapons*, The Oxford Martin policy Paper, 2014, p. 32.

Lewis John, The Case for Regulating Fully Autonomous Weapons, *The Yale Law Journal*, 2015, Vol. 124, p. 1309-1325.

Libel Tamir and Emily Boulter, Unmanned Aerial vehicles in the Israel Defence Forces: A Precursor to a Military Robotic revolution? *The RUSI Journal*, April/May 2015, vol. 160, No. 2, p. 68-71.

Libicki Martin C. 2009. *Cyber Deterrence and Cyber War*, RAND Corporation.

Lieblich Eliary and Eyal Benvenisti, The Obligation to Exercise Discretion in Warfare: Why Autonomous Weapon Systems are Unlawful, Global Trust Working Paper Series 10/2014.

Liivoja, Rain and McCormack, Tim, Law in the Virtual Battlespace: The Tallinn Manual and the *Jus in Bello*, University of Melbourne Legal Studies Research Paper No. 650, 2013.

Liivoja, Rain, Chivalry without a Horse: Military Honour and the Modern Law of Armed Conflict, ENDC Proceedings, Volume 15, 2012, p. 75-100.

Lim Zhifeng, The Rise of Robots and the Implications for Military Organizations, Unpublished Thesis, Naval Postgraduate School California, 2013, p. 1-86.

Lin Herbert, Cyber Conflict and International Humanitarian Law, *International Review of the Red Cross*, Vol. 94, No. 886, Summer 2012, p. 515-532.

Lin Patrick, George Bekey and Keith Abney, Autonomous Military Robotics: Risk, Ethics, and Design, Report prepared for the US Department of Navy, Office of Naval Research, 2008, p. 112.

Lin Patrick, The right to life and the Martens Clause, paper Presented at: Convention on Certain Conventional Weapons (CCW) meeting of experts on lethal autonomous weapons systems (LAWS), at United Nations in Geneva, Switzerland on 13-17 April 2015.

Liu Hin-Yan, Categorization and legality of autonomous and remote weapons systems, *International Review of the Red Cross*, Vol. 94, No. 886, Summer 2012, p. 627-652.

Losing Humanity: The Case Against Killer Robots, Human Rights Watch, November 2012.

Lucas George R, Automated Warfare, *Stanford Law and Policy Review*, Vol. 25, 2014, p. 317-339.

Lucas George R., Legal and Ethical Percepts Governing Emerging Military Technologies: Research and Use, *Utah Law Review*, 2013, No. 5, p. 1271-1281.

Lucas George R., Legal and Ethical Precepts Governing Emerging Military Technologies: Research and Use, *Amsterdam Law Forum*, Vol. 6, No. 1, 2014, p. 23-34.

Lulf Charlotte, Modern technologies and Targeting Under International Humanitarian Law, IEHV Working Paper, Vol. 3, No. 3, December 2013, p. 1-63.

Makin N. S., Future warfare or Future Folly? Autonomous Weapon Systems on the Future Battlefield: An Assessment of Ethical and Legal Implications in their Potential Use, April 2008, Master of Defence Studies Research Project, Canadian Forces College, JSCP 34.

March of the Robots, *The Economist*, Technology Quarterly: Q2, 2012, p. 11-12.

Marra William C., and Sonia K. McNeil, Understanding "The Loop": Regulating the Next Generation of War Machines, *Harvard Journal of Law & Public Policy*, Vol. 36, No. 3, 2013, p. 1139-1185.

Marsh Nicholas, Defining the Scope of Autonomy: Issues for the Campaign to Stop Killer Robots, Peace Research Institute Oslo (PRIO) Policy Brief 2, 2014.

Mathias Becca, Autonomous Battlefield Robots, Position paper 9, available at: https://www.sandiego.edu/cas/documents/ma-international-relations/PositionPaper-BeccaMathias.pdf.

Matthews William, Robot or Not? Army Dismisses Completely Soldierless Battlefield, Army, November 2014, p. 35-38.

McDaniel Erin A., Robot Wars: Legal, Ethical Dilemmas of Using Unmanned Robotic Systems in 21st Century Warfare and Beyond, Master's Thesis submitted to Missouri State University, 2008.

McFarland Tim and Tim McCormack, Mind the Gap: Can Developers of Autonomous Weapons Systems be Liable for War Crimes? Vol. 90, *International Law Studies*, 2014, p. 361-385.

McGlynn Daniel, Robotic Warfare: Should Autonomous Military Weapons be Banned, January 2015, available at: http://library.cqpress.com/cqresearcher/document.php?id=cqresrre2015012300.

McLaughlin Robert, 'Unmanned Naval Vehicles and the Law of Naval Warfare', in Nasu Hitoshi and Robert McLaughlin (ed.). 2014. *New Technologies and the Law of Armed Conflict*, The Hague: TMC Asser Press, p. 229-246.

Melzer Nils, *Human rights implications of the usage of drones and unmanned robots in warfare*, Directorate-General for the External Policies of the Union, European Parliament's Subcommittee on Human Rights, 2013, p. 53.

Merchant, Gary E. et. al., International Governance of Autonomous Military Robots, *The Columbia Science and Technology Law Review*, Vol. XII, 2011, p. 271-315.

Meron, T., The Martens Clause, Principle of Humanity, and Dictates of the Public Conscience, American Journal of International Law, Vol. 94, 2000. p. 78-89.

Mies Gerald, Military robots of the present and the future, *AARMS*, Vol. 9, No. 1, 2010, p. 125-137.

Mind the Gap: The Lack of Accountability for Killer Robots, Human Rights Watch, p. 44, April 2015.

Moor James H., Taking the Intentional Stance toward Robot Ethics, APA Newsletters, Vol. 6, No. 2, Spring 2007.

Moor James H., The Nature, Importance, and Difficulty of Machine Ethics, IEEE Intelligent Systems, July/August 2006, p. 18-21.

Moshkina Lilia and Arkin Ronald C., *Lethality and Autonomous Systems: Survey Design and Results*, Technical Report, 2008, The US Army Research Office.

Mullner Karl, Remote-Controlled Aerial Vehicles – Made-to-Measure Effectiveness for Protection of our Soldiers on Mission, *Ethics and the Armed Forces*, Issue 2014/1, p. 23-30.

Muller Vincent C. and Simpson Thomas W., Killer robots: Regulate, don't ban, The BSG Policy Memo, November 2014.

Muller Vincent C. and Thomas Simpson W., Autonomous Killer Robots

Are Probably Good News, pp.16, available at: http://www.bsg.ox.ac.uk/sites/www.bsg.ox.ac.uk/files/2014-Killer-Robots-Policy-Paper.pdf.

Nasu Hitoshi and Robert McLaughlin (ed.). 2014. *New Technologies and the Law of Armed Conflict*, The Hague: TMC Asser Press.

Naval Aviation Vision 2014-2025, Naval Aviation Enterprise, The US Navy.

New Roles for Technology: Rise of the robots, *The Economist*, 29 March 2014.

Newton Michael A., Back to the Future: Reflections on the Advent of Autonomous Weapons Systems, *Case W. Res. J. Int'l L.*, Vol. 47, No. 1, 2015, p. 6-23.

No Man's Land: Tech Considerations for Canada's Future Army, Canadian Army Land Warfare Centre Kingston, Ontario, 2014, p. 211.

Noone Gregory P. and Noone Diana C., The Debate Over Autonomous Weapons Systems, *Case W. Res. J. Int'l L.*, Vol. 47, No. 1, 2015, p. 25-35.

Noorman Merel and Deborah G. Johnson, Negotiating autonomy and responsibility in military robots, *Ethics Inf Technol*, Vol. 16, 2014, p. 51–62.

Plaw Avery, Upholding the Principle of Distinction in Counter-Terrorist Operations: A Dialogue, *Journal of Military Ethics*, Vol. 9, No. 1, 2010, p. 3-22.

Olsthoorn Peter, Risks and Robots-some ethical issues, available at: http://ssrn.com/abstract=2436309.

O'Connell Mary Ellen, Banning Autonomous Killing, Notre Dame Law School Legal Studies Research Paper No. 1445.

O'Gorman Rob and Chris Abbott, Remote Control War: Unmanned combat air vehicles in China, India, Israel, Iran, Russia and Turkey, September 2013; Open Briefing, London.

Ohlin Jens David, The Combatant's Stance: Autonomous Weapons on the Battlefield, *International Law Studies*, Vol. 92, 2016, p. 1-30.

O'Meara, Richard M., The Rules of War and the Use of Unarmed,

Remotely Operated, and Autonomous Robotics Systems, Platforms and Weapons…Some Cautions, available at: http://robots.law.miami.edu/wp-content/uploads/2012/01/Omeara-intersection.pdf .

O'Meara, Richard M., Emerging Military Technologies in the 21st Century: Assessing the Need for Governance, Unpublished Thesis, 2011.

Parker Sarah and Katherine Green, *A Decade of Implementing the United Nations Programme of Action on Small Arms and Light Weapons: Analysis of National Reports*, United Nations Institute for Disarmament Research (UNIDIR), Geneva, Switzerland, 2012.

Pau Mauro, Robots in warfare: Why we need to start an ethical debate, available at: http://folk.uio.no/mauro/papers/robowars.pdf.

Pradhan Sushil, Robotics in Warfare, *Journal of the United Service Institution of India*, Vol. CXLI, No. 588, April-June 2012.

Quintana Elizabeth, The Ethics and Legal Implications of Military Unmanned Vehicles, Occasional Paper, *RUSI*, p. 22.

Radziejowska Maria, Remote and Autonomous: From Drones to "Killer Robots", *Strategic File*, No. 24 (60), October 2014, The Polish Institute of International Affairs, p. 1-6.

Rappert Brian, Richard Moyes, Anna Crowe and Thomas Nash, The roles of civil society in the development of standards around new weapons and other technologies of warfare, *International Review of the Red Cross*, Vol. 94, No. 886, Summer 2012, p. 765-786.

Reed John, Beyond Drones: The Next-Generation of Autonomous Weapons Cannot be Developed in Secrecy, available at: http://justsecurity.org/20825/autonomous-weapons-developed-secrecy/.

Reeves Shane R. and Jeffery S. Thurnher, Are We Reaching a Tipping Point? How Contemporary Challenges Are Affecting the Military Necessity-Humanity Balance, *Harvard National Security Journal Features*, 2013, p. 1-12.

Reeves R. and William J. Johnson, Autonomous Weapons: Are You Sure These Are Killer Robots? Can We Talk About It? *The Army Lawyer*, April 2014, p. 25-31.

Regional Perspectives on Norms of Behaviour for Outer Space Activities, United Nations Institute for Disarmament Research, Geneva, 2015.

Report of the ICRC Expert Meeting on 'Autonomous weapon systems: technical, military, legal and humanitarian aspects', 26-28 March 2014, International Committee of the Red Cross, Geneva.

Review of the 2012 US Policy on Autonomy in Weapons Systems, Human Rights Watch and Harvard Law School International Human Rights Clinic, April 2013, p. 1-9.

Rhodes Bill. 2009. An *Introduction to Military Ethics: A Reference Handbook*, Praeger Security International.

ROBOLAW, Regulating Emerging Robotic Technologies in Europe: Robotics facing Law and Ethics, Guidelines on Regulating Robotics, D6.2.

Robotics and Autonomous Systems Industry, *Industry Study*, Spring 2011, The Industrial College of the Armed Forces, National Defence University, Washington, DC.

Robotics Strategy: White Paper, The US Department of Army, 19 March 2009.

Roff Heather M., Lethal Autonomous Weapons and Jus Ad Bellum Proportionality, *Case W. Res. J. Int'l L.*, Vol. 47, No. 1, 2015, p. 37-52.

Roff, Heather M., The Strategic Robot Problem: Lethal Autonomous Weapons in War, *Journal of Military Ethics*, Vol. 13, No. 3, 2014, p. 211-227..

Rogers Jay Logan, Legal Judgment Day for the Rise of the Machine: A National Approach to regulating Fully Autonomous Weapons, *Arizona Law Review*, Vol. 56, No. 4, 2014, p. 1257-1272.

Royakkers Lamber and Rinie van Est, The cubicle warrior: the marionette of digitalized warfare, *Ethics Inf Technol*, Vol. 12, 2010, p. 289–296.

Rus Daniela, How Technological Breakthroughs Will Transform Everyday Life, *Foreign Affairs*, Vol. 94, No. 4, July/August 2015, p 226.

Sapaty Peter Simon, Military Robotics: Latest Trends and Spatial Grasp

Solutions, *International Journal of Advanced Research in Artificial Intelligence*, Vol. 4, No.4, 2015, p. 9-18.

Sartor Giovanni, The Autonomy of Automated Weapons, Ludovica Glorioso, Anna-Maria Osula (Eds.), Proceedings of First Workshop on Ethics of Cyber Conflict, Tallinn 2014, p. 73-81.

Sassoli Marco, Autonomous Weapons and International Humanitarian Law: Advantages, Open Technical Questions and Legal Issues to be Clarified, Vol. 90, *International Law Studies*, 2014, p. 308-339.

Sassoli Marco, Autonomous Weapons – Potential Advantages for the Respect of International Humanitarian Law, PHAP, 2 March 2013.

Sauer Frank, Autonomous Weapons Systems: Humanising or Dehumanising Warfare, Global Governance Spotlight, 4/2014, Germany: Development and Peace Foundation.

Scharre Paul, Autonomy, 'Killer Robots,' and Human Control in the Use of Force – Part I and II, *Just Security*, 9 July 2014.

Scharre Paul, Robotics on the Battlefield Part I: Range, Persistence and Daring, May 2014, Centre for a New American Security.

Scharre Paul, Robotics on the Battlefield Part II: The Coming Swarm, October 2014, Centre for a New American Security.

Scharre Paul, Horowitz Machael C and Kalley Sayler, Autonomous Weapons at the UN: A Primer for delegates, Centre for New American Security, April 2015.

Schmitt, Michael N., Autonomous Weapon Systems and International Humanitarian Law: A Reply to the Critics, *Harvard National Security Journal Features,* 2013, p. 1-37.

Schmitt, Michael N., The Principle of Discrimination in 21st Century Warfare, *Yale Human Rights and Development Journal*, Vol. 2, Issue 1, 1999, p. 143-182.

Schmitt, Michael N., International Law and Military Operations in Space, *Max Planck UNYB*, Vol. 10, 2006, p. 89-125.

Schmitt, Michael N. and Jeffrey S. Thurnher, Out of the Loop: Autonomous

Weapon Systems and the Law of Armed Conflict, *Harvard National Security Journal*, Vol. 4, 2013, p. 231-281.

Schmitt, Michael N., Charting the Legal Geography of Non-International Armed Conflict, Vol. 90, *International Law Studies*, p. 1-19, 2014.

Schmitt, Michael N., War Technology and the Law of Armed Conflict, *International Law Studies*, Vol. 82, p. 137-182.

Schmitt, Michael N., *War Technology and the International Humanitarian Law*, Occasional Paper 2005, HPCR Harvard University.

Schmitt, Michael N., The Law for Cyber Targeting, The Tallin Paper No.7, 2015.

Schornig Niklas, Robot warriors: why the Western investment into military robots might backfire, Peace Research Institute Frankfurt-Report No. 100.

Shaking the Foundations: The Human Rights Implications of Killer Robots, Human Rights Watch Report, 2014, p. 33.

Sharkey Noel, Automating Warfare: lessons learned from the drones, *Journal of Law, Information and Science*, Vol. 21 (2), 2012, p. 8.

Sharkey Noel, Grounds for Discrimination: Autonomous Robot Weapons, *RUSI Defence Systems*, October 2008, p. 86-89.

Sharkey Noel, Towards a principle for the human supervisory control of robot weapons, 2014, available at: https://www.mini-symposium-tokyo. info/ICRA2014/sharkey2014.pdf, accessed 7 July 2015, p. 16.

Sharkey, Noel. 2012. 'Killing made easy: From joysticks to politics.' In *Robot Ethics: The Ethical and Social Implications of Robotics*, Lin, Patrick, Keith Abney, and George Bekey (ed.), Cambridge, Mass: MIT Press, p. 111–128.

Sharkey Noel E., The Evitability of Autonomous Robot Warfare, *International Review of the Red Cross*, Vol. 94, No. 886, Summer 2012, p. 787-799.

Siboni Gabi and Yoni Eshpar, Dilemmas in the Use of Autonomous Weapons, *Strategic Assessment*, Vol. 16, No. 4, January 2014, p. 75-87.

Siddiqui Huma, Keeping an Eye, Underwater, *The Financial Express*, 6 April 2015.

Simpson Thomas W. and Muller Vincent C., Just War and Robots Killing, *The Philosophical Quarterly*, August 2015, available at: http://pq.oxfordjournals.org/.

Singer, P. W. 2009. *Wired for War: The Robotics Revolution and Conflict in the 21st Century*, New York: Penguin books.

Singer P.W., Robots at War: The New Battlefield, *Wilson Quarterly*, Autumn 2008, p. 30-48.

Singer P.W., Military Robots and the Laws of War, The New Atlantis, 2009, p. 25-45.

Singh Harbhajan, Cyber Warfare: Dangerous Trends, *Journal of the USI*, Vol. CXLI. No. 584, April-June 2011, p. 228-235.

Sloan Robert D., Puzzles of Proportion and the "Reasonable Military Commander": Reflections on the Law, Ethics, and Geopolitics of Proportionality, *Harvard National Security Journal*, Vol. 6, 2015, p. 299-343.

Solomon Cara, The Human Rights Implications of Killer Robots, 16 June 2014, available at: http://hrp.law.harvard.edu/staff/the-human-rights-implications-of-killer-robots/.

Sparrow, R., Killer Robot, *Journal of Applied Philosophy*, Vol. 24(1), March 2007, p. 62-77.

Sparrow, R., Predators or Plowshares: Arms Control of Robotic Weapons, *IEEE Tech and Society Magazine*, Spring 2009, p. 25-29.

Sparrow Robert, The Ethical Challenges of Military Robots, p. 16, available at: http://www.bundesheer.at/pdf_pool/publikationen/20101105_et_ethical_and_legal_aspects_of_unmanned_systems_sparrow.pdf .

Sparrow, R. 2011. 'Robotic Weapons and the Future of War', in Jessica Wolfendale and Paolo Tripodi (eds.), *New Wars and New Soldiers: Military Ethics in the Contemporary World*, Surrey, Burlington, VT: Ashgate, p. 117-133.

Sparrow Robert, Twenty Seconds to Comply: Autonomous Weapon Systems and the Recognition of Surrender, *International Law Studies*, Vol. 91, 2015, p. 299-728.

Statman Daniel, Drones, Robots and the Ethics of War, *Ethics and the Armed Forces*, Issue 2014/1, p. 41-45.

Stewart Darren M., New Technology and the Law of Armed Conflict, *International Law Studies*, Vol. 87, 2009, p. 271-298.

Sukman Daniel, Lethal Autonomous Systems and the Future of Warfare, *Canadian Military Journal*, Vol. 16, No 1, Winter 2015.

Sullins John P., An Ethical Analysis of the Case for Robotic Weapons Arms Control, Paper Submitted at 5th International Conference on Cyber Conflict, pp. 1-20.

Sullins John P., The Ambiguous Ethical Status of Autonomous Robots, available at: https://www.aaai.org/Papers/Symposia/Fall/2005/FS-05-06/FS05-06-019.pdf.

Technological Challenges for the Humanitarian Legal Framework, *Proceedings of the 11th Bruges Colloquium*, 21-22 October 2010.

Terzian Dan, The Right to Bear (Robotic) Arms, *Penn State Law Review*, Vol. 117, No. 3, 2013, p. 755-796.

The Guardian view on robots as weapons: the human factor, Editorial, 13 April 2015.

The Military Commander and the Law, The Judge Advocate General's School, Maxwell Air Force Base, Alabama, USA, 2014.

The United States, Department of Defence, Law of War Manual, 2015.

The Weaponization of Increasingly Autonomous Technologies: Considering Ethics and Social Values, The United Nations Institute for Disarmament Research (UNIDIR), 2015, pp. 14.

The Weaponization of Increasingly Autonomous Technologies: Considering how Meaningful Human Control might move the discussion forward, UNIDIR, 2014, p. 12.

The Weaponization of Increasing Autonomous Technologies in the Maritime Environment: Testing the Waters, UNIDIR, No. 4, UNIDIR Resources, 2015.

Thomas, Bradan T., Autonomous weapon Systems: The Anatomy of Autonomy and the Legality of Lethality, *Houston Journal of International Law*, Vol. 37, No. 1, 2015, p. 235-274.

Thurnher Jeffery S., No One at Control: Legal Implications of Autonomous Targeting, *Joint Force Quarterly*, Issue 67, No. 4, 2012, p. 77-84.

Thurnher Jeffery S., The Law That Applies to Autonomous Weapon Systems, *Insight*, Vol. 17, No. 4, January 2013, American Society of International Law.

Thurnher Jeffery S., 'Examining Autonomous Weapon Systems from a Law of Armed Conflict Perspective', in Nasu Hitoshi and Robert McLaughlin (ed.). 2014. *New Technologies and the Law of Armed Conflict*, The Hague: TMC Asser Press, p. 213-228.

Toscano, Christopher P., "Friend of Humans": An Argument for Developing Autonomous Weapons Systems, *Journal of National Security and Policy*, 6 May 2015.

Tripodi Paolo and Jessica Wolfendale. 2011. *New Wars and New Soldiers: Military Ethics in the Contemporary World*, UK: Ashgate.

Trneny Michal, State of the art military robots and future of warfare, 2013, available at: http://www.fi.muni.cz/usr/gruska/future13/esseytrneny.pdf.

UN Security Council: Report of the Secretary-General on the protection of civilians in armed conflict, S/2013/689, 22 November 2013.

Unmanned Systems Integrated Roadmap: FY 2009-2034, The US Department of Defence, p. 210.

Vallejo Daniel, Electric Currents: Programming Legal Status into Autonomous Unmanned Maritime Vehicles, *Case W. Res. J. Int'l L.* , Vol. 47, (2015), p. 405-428.

Vallor Shannon, The Future of Military Virtue: Autonomous Systems and the Moral Disliking of the Military, Paper Presented at 2013, 5th International Conference on Cyber Conflict.

Varga Col. Attila Ferenc, Rules of Engagement in Robotic Warfare, *Vojenske Reflexive*, 2013 p. 63-78.

Velez-Green Alexander, The Foreign Policy Essay: The South Korean Sentry—A "Killer Robot" to Prevent War, 1 March 2015, The Lawfare Institute.

Vilmer Jean-Baptiste Jeangene, Terminator Ethics: Should We Ban "Killer Robots" *Ethics and International Affairs*, 23 March 2015.

Voort Marlies Van de, Wolter Pieters and Luca Consoli, Refining the ethics of computer-made decisions: a classification of moral mediation by ubiquitous machines, *Ethics Inf Technol*, Vol. 17, 2015 p. 41-56.

Wagner Markus, The Dehumanization of International Humanitarian Law: Legal, Ethical, and Political Implications of Autonomous Weapon Systems, *Vanderbilt Journal of Transnational Law*, Vol. 47, 2014, p. 1371-1424.

Wallach Wendell and Colin Allen, Framing robot arms control, *Ethics Inf Technol*, Vol. 15, 2013, p. 125–135.

Waser Mark R., The Bright Red Line of Responsibility, available at: http://www.iacap.org/proceedings_IACAP13/paper_29.pdf.

Weaver John Frank, Asimov's Three Laws Are Not an International Treaty: How to make treaties govern "killer robots", 1 December 2014, The Citizens Guide to the Future.

Weber Jutta, Robotic Warfare, Human Rights and the Rhetoric of Ethical Machines, *Ethics and Robotics*, 2009, p. 83-103.

Weiner Tim, New Model Army Soldier Rolls Closer to Battle, 16 February 2015, available at: http://www.nytimes.com/2005/02/16/technology/new-model-army-soldierrolls-closer-to-battle.html.

Weizmann Nathalie, *Autonomous Weapon Systems under International Law*, Geneva Academy of International Humanitarian Law and Human Rights, November 2014, p. 27.

Wellbrink Jorg, My New Fellow Soldier – Corporal Robot? *Ethics and Armed Forces*, Vol. 2014, No. 1, p. 46-49.

White Andrew, Maritime remotely Operated Vehicles (ROV) and Autonomous Underwater Vehicles (AUV), *Military Technology*, June 2015, p. 86-88.

Williamson Richard L., Hard Law, Soft Law, and Non-Law in Multilateral Arms Control: Some Compliance Hypotheses, *Chicago Journal of International Law*, Vol. 4: No. 1, 2003, p. 59-82.

Winslow Lance, Unmanned Vehicle: Robotic Warfare "Hide and seek Strategies", 2007, available at www.worldthinktank.net.

Work Robert O. and Shawn Brimley, *20YY: Preparing for War in the Robotic Age*, Centre for a New American Security, January 2014, p. 1-40.

Yde Iben, The push towards autonomy: an insight into the legal implications of self-targeting weapon systems, Aarhus University, Department of Law, Royal Danish Defence College, 2014, p. 19.

Zawieska Karolina, Do robots equal humans? Anthropomorphic terminology in LAWS, Industrial Research Institute for Automation and Measurements PIAP, Warsaw, Poland, p. 1-3.

Source of Photographs

(Chapter 3)

Dragon Robot: http://www.spacedaily.com/news/robot-04r.html.

Guardium: https://defense-update.com/products/g/guardium.htm.

irobot 510 Packbot: http://www.irobot.com/~/media/Files/Robots/Defense/PackBot/iRobot-510-PackBot-Specs.pdf.

BigDog/MULE: http://www.bostondynamics.com/robot_bigdog.html.

MAARS: https://www.qinetiq-na.com/products/unmanned-systems/maars/.

SGR-A1: http://www.ubergizmo.com/2014/09/samsung-sgr-a1-robot-sentry-is-one-cold-machine/.

DRDO's Daksh: http://aermech.in/drdo-daksh-rov-remotely-operated-vehiclefirst-robot-to-defuse-bombs/.

GRUNT: http://www.kvh.com/Press-Room/Press-Release-Library/2009/090224-Frontline-Robotics-Using-CNS5000.aspx.

Platform-M: https://img.rt.com/files/2015.07/original/55ad03eac36188c034 8b4571.jpg.

Taranis: http://www.hngn.com/articles/135102/20150930/taranis-drone-bae-systems-creates-u-k-s-advanced-secretive.htm.

X- 47B: http://www.popularmechanics.com/military/research/a15230/x-47b-mid-air-refueling/.

Iron Dome Battery: http://www.jewishvirtuallibrary.org/jsource/Peace/IronDome.html.

AWS-CTUV: http://www.darpa.mil/program/anti-submarine-warfare-continuous-trail-unmanned-vessel.

ONR LDUUV Prototype: http://www.navaldrones.com/LDUUV-INP.html.

Littoral Battlesapce Glider: http://www.navaldrones.com/LBS-Glider.html.

MK 15 Phalanx:http://www.raytheon.com/capabilities/products/phalanx/.

DRDO's AUV: http://www.indiandefensenews.in/2015/02/indian-navy-impressed-with-drdo-s.html.

Index

A

AEGIS System 58

Anthrax 2

Antipersonnel Landmines Treaty 187

Anti-Submarine Warfare (AWS) 55

Arms Trade Treaty 8, 70, 82-85, 93, 99, 100, 195, 208, 218

Australian Defence Force 128, 129

Autonomous Aerial Refuelling 48

Autonomous robots 2

Autonomous underwater vehicles 53, 59

Autonomous Weapons Systems 55, 95, 96, 98-102, 118, 190, 217, 218, 222-224, 227, 230, 232, 235, 239

Future of 6, 118, 140, 193, 226, 228, 237-239

B

Battle Management and Weapon Control 49

Bharat Electronics Ltd 43

BigDog 36

Bionic Hornets 52

Brimstone 51

C

CCW Meeting of Experts on LAWS 149, 161

Christof Heyns 5, 22, 89, 101, 102, 148, 173, 181

Close-In System 58

Cluster munitions 12, 69, 97, 124, 187, 189, 193

Convention on Cluster Munitions 7, 187, 193

Crusher 41

D

Daksh 43

Defence Advanced Research Projects Agency 55

Defence Research and Development Organization 3, 43, 44, 60, 66

Demilitarized Zone 2, 3, 41, 120

Drone 15, 16, 19, 56, 183, 190, 191

Dum-Dum bullets 124, 141

E

Electronic counter-countermeasure 59

Electronic countermeasure 59

Electronic support measure 59

Ethical LAWS 111

Expanding bullets 69, 124

Explosive Ordnance Disposal 30, 31, 64

Explosive Remnants of War 12

F

Fire and forget 18

Forward-Looking InfraRed 41

Fully autonomous weapons 88, 89

G

Geneva Convention 91, 103, 197, 200

Geneva Conventions of 1949 71, 78, 83, 106, 185, 192, 195, 197, 208, 209

Ground-based Automated Systems 31

Crusher 41

Daksh 43

GRUNT 44

Guardium 33

iRobot UGVs 35

MAARS 39

Platform-M 45

SGR-A1 41, 42

SWORDS 34

TALON 31, 32, 34, 35, 64, 65, 84

GRUNT 44

Guardium 33, 34, 76, 101, 177

H

Hague Convention 124, 138, 141, 147, 195

Human-out-of-the-loop weapon 20

Human Rights Committee 86, 87, 90, 103

I

Improvised Explosive Devices x, 35, 64, 100

Intelligence, Surveillance and Reconnaissance 30, 54, 56, 67

International Committee for Robot Arms Control 6, 168, 174, 175, 176, 180

International Committee of the Red Cross 6, 24, 25, 29, 75, 78, 95, 96, 97, 113, 118, 126, 137, 139, 140, 142, 145, 147, 148, 160, 161, 170, 171, 173, 174, 176, 184-186, 192, 193, 217, 234

International Court of Justice 101, 125, 138, 141, 146, 162

International Covenant on Civil and Political Rights 86, 89, 91, 92, 101, 102, 103

International Criminal Court 74, 78, 92, 96, 124, 141, 207

International Humanitarian Law 3, 5-6, 6, 9-12, 24, 29, 53, 55, 62, 69-71, 73, 75-81, 83, 86, 92-98, 100, 119, 122, 124-125, 127, 130, 132-135, 139, 143, 145-146, 149-150, 152-158, 160-164, 166-173, 175, 178, 180-182, 185-192, 195, 200, 208, 210, 215-216, 218, 221, 230, 235-236, 240

iRobot 710 Kobra 36

iRobot UGVs 35

iRobot 510 PackBot 35, 36

Iron Dome 15, 28, 49, 50, 51, 63, 66, 105

Israel Aerospace Industries Limited 3

J

Jus ad bellum 69, 70, 93, 220

K

Korean Robotic Sentry

SGR-A1 28, 41, 42, 65, 76, 114, 177

Kosovo conflict 109

L

Landmines 12, 24, 69, 124, 168, 187, 193

Large Displacement Unmanned Undersea Vehicle 56

Law of Armed Conflict 117, 129, 140, 142-144, 146, 147, 193, 216, 219-221, 225-227, 229, 231, 232, 236, 238, 239

Legal Review of Weapons 123

Legged Squat Support System x

Lethal Autonomous Robot x, 22

Lethal Autonomous Weapon Systems (LAWS) 3-9, 13, 18, 26, 53, 63-64, 69-81, 84, 86, 90-93, 101, 105-106, 108-116, 122-123, 125-126, 131, 135, 137-139, 143, 148-173, 176-195, 212, 229, 241

Levels of Autonomy 13, 14, 15

Long-Range Anti-Ship Missile 58, 59

M

Maritime Autonomous Weapon Systems 53

Problems with 62

Meaningful Human Control 149, 150, 153, 154, 159, 163, 164, 166, 169, 170, 172, 226, 238

Military robots 108, 152, 168, 232, 236, 239

MK 15 Phalanx Close-In Weapons System 3

Modular Advanced Armed Robotic System 32, 39, 40

N

NBS MANTIS 52

Non-governmental organization x, 175

Non-lethal incapacitating weapons 87

Non-state actors 9, 85, 92, 166, 168, 181

P

Peace Research Institute Oslo 51, 66, 230

Permanent Court of International Justice 78

PHALANX system 177

Platform-M 45

Protocol I Additional to the Geneva Conventions of 12 August 1949. 96,106,142,192, 200

Protocol VI on LAWS (Draft) 9, 189, 212-213

R

Reconnaissance, surveillance and target acquisition 39

Remotely operated vehicles 34, 59, 67, 241

Robotics 2, 11, 32, 44, 64, 65, 104, 107, 117, 118, 119, 180, 215, 216, 218-220, 222-224, 226, 227, 229, 233-237, 240

Robotic Weapon Systems 3, 225

Rome Statute of the International Criminal Court 74, 78, 92, 96, 124, 207

Rules of Engagement 39, 81, 112, 114, 140, 145, 180, 240

S

Semi-autonomous Weapon Systems 17, 18

SGR-A1 3, 24, 28, 41, 42, 65, 76, 101, 105, 114, 120, 177, 242, 244-246

Special Weapons Observation Remote Direct-Action System 34, 35, 39, 65, 84, 227

Stockholm International Peace Research Institute 99, 143, 146, 147, 169, 175, 216, 219

St. Petersburg Declaration 195

SWARM 52, 177

SWORDS 34, 35, 39, 65, 84, 227

T

TALON 31, 32, 34, 35, 64, 65, 84

Taranis aircraft 3, 19, 46, 47

TomCar 33

Treaty of Versailles 124

U

UN Human Rights Council 7, 91, 148

United Nations 6, 67, 68, 82, 83, 97, 99, 102, 148, 152, 161, 162, 176, 188, 208, 209, 222, 229, 233, 234, 238

Universal Declaration of Human Rights 86, 90, 91, 101, 102, 103, 120, 208

Unmanned Aerial Systems

Taranis 46

X-47 systems 46

Unmanned Aerial Vehicle 46, 59

Unmanned aircraft systems 30, 59, 68

Unmanned Combat Aerial Vehicle 47, 48, 52, 137

Unmanned Ground System 30

Unmanned Ground Vehicle 33, 39, 41, 44, 65

Unmanned Surface Vehicle 67, 68

Unmanned System 53

Unmanned Underwater Vehicle 31, 67

UN Register of Conventional Arms 82, 93

US Army's Training and Doctrine Command 10

X

X-47B autonomous aircraft 3, 47, 48, 49, 65, 76, 177